Estimation of Parameters for Animal Populations

∞

Estimation of Parameters for Animal Populations

∞

A PRIMER
FOR THE REST OF US

∞

LARKIN A. POWELL
AND
GEORGE A. GALE

Caught Napping Publications
Lincoln, Nebraska USA

Caught Napping Publications
1120 S. 40th Street
Lincoln, NE 68510 USA

Copyright © 2015 by Larkin A. Powell and George A. Gale

*All rights reserved,
including the rights of reproduction
in whole or in part in any form.
Please contact publisher for permission.*

ISBN 978-1-329-06151-4

Cover photo: Kelly Turek (*used with permission*), northern pike (*Esox lucius*), Nebraska, USA
Back cover photo: Dusit Ngoprasert (*used with permission*), Asiatic black bears (*Ursus thibetanus*),
Thailand

∞

TO

DR. JAMES D. NICHOLS

He inspired us with his ability and desire to communicate complex, quantitative concepts to any person.

CONTENTS

Preface	ix
Foreword	xv

Section I – Background
    1-Models of Populations   1
    2-Context and Inference   11
    3-Maximum Likelihood Estimation   21
    4-AIC and Model Selection   39
    5-Variance and the Delta Method   61
    6-Linear Models   71
    7-Introduction to Mark-Recapture Models   83

Section II – Mark-recapture

    8-Estimation of Population Size: Closed Populations   91
    9-Known Fate   103
    10-Cormack-Jolly-Seber: Apparent Survival   123
    11-Recovery Data   139
    12-Multi-state Models   147
    13-Robust Design   157

Section III – Survey data
    14-Introduction to Survey Sampling   169
    15-Occupancy Modeling   185
    16-Double-observer Methods   195
    17-Removal Sampling   205
    18-N-mixture Models   213
    19-Distance Sampling   223

Preface

"Education is the kindling of a flame, not the filling of a vessel."
-- attributed to Socrates

Our goal for this primer

The reader, or potential reader, of this book may wish to know the goals behind the creation and publication of the materials presented here. Our goal is simple: we hope to present information to beginning learners in the field of demographic parameter estimation in a manner that allows a firm foundation from which to expand throughout your career. A secondary goal is that we wish to present the information in an easily accessible and inexpensive format.

We believe that two gaps exist, currently, in the types of books available in this field of study. First, many good books about parameter estimation are very expensive, and this limits the study of wildlife demography in many countries in our world. Those countries, ironically, have the highest need for scientists who can contribute to local conservation efforts. So, we present this book as downloadable PDFs and as a relatively inexpensive hardcopy book that is priced with no profit margin. The book costs what it costs to produce it. We hope that the accessibility helps.

Second, we have aimed our information to the beginning student, especially students with limited mathematical backgrounds. We present the basic ideas and logic behind the models for estimation of demographic parameters. Such information is often skipped quickly in some other books in favor of in-depth exploration of higher-order mathematics or statistics.

The information presented here may suffice, for some learners, to provide an understanding of the models being used by colleagues. For example, if you work in the a government ministry for wildlife in a country in southern Africa and you are collaborating on a study of survival of black rhinos based on radio-telemetry data, you may not need to know how to do the analysis—but it would sure be nice to understand what your colleagues are talking about. This book is for you.

For other learners, this book will be their first book, but it should not be their last. Perhaps you are an undergraduate student who needs a solid foundation at the undergraduate level, or perhaps you are a graduate student who wishes to conduct rigorous analyses of data. You will find, elsewhere, excellent expansions on the introductory information we present. In fact, at the end of each chapter and at the end of this forward, we list a set of resources that represent the 'next step' in your journey towards understanding more complex models and exploration of further depth in the basic models we cover in this text. But, we believe that the journey to rigorous analyses starts with the "kindling of the flame" that requires a good understanding of why these models work and why they are constructed in a particular manner.

x

How to read and use this primer

We have designed the content of the primer with a purpose. Everyone should read Section I, which provides an overview of basic concepts that are common to analyses of captured and marked animals as well as animals observed in surveys. We place the mark-recapture chapters in Section II for two reasons: (1) historically, quantitative methods in wildlife ecology and management were developed for marked animals before rigorous methods for analysis of survey data were developed—so mark-recapture has historical precedence that seems proper to carry through in this primer, and (2) many of the methods for analysis of survey data in Section III rely on an understanding of encounter probabilities. The reader will be very familiar with encounter probabilities after working through Section II.

Study Goals: "I want to estimate…"					
Study Type	Presence, absence	Pop. size, abundance	Density	Survival	Survival, movement
Count survey	Ch. 15: Occupancy	Ch. 16: Double observer	Ch. 19: Distance sampling		
		Ch. 17: Removal			
		Ch. 18: N-mixture			
Mark-recapture		Ch. 8: Closed mark-recapture		Ch. 9: Known-fate	Ch. 12: Multi-state
				Ch. 10: C-J-S	Ch. 13: Robust design
				Ch. 11: Recovery	
Which method should I use?					

When your authors took our first courses in parameter estimation, the courses consisted of material covered in Chapters 1-13 (there was no widespread use of robust design in the mid-1990s!) and the distance-based survey methods found in Chapter 20. Since our first courses, some smart folks have created new methods (e.g., robust design, N-mixture models, occupancy modeling). And, even more people have started to modify these methods in creative ways—we now have options for multi-state robust design analyses, or you can find "open" N-mixture models. Occupancy models have been modified to assess the co-occurrence of species. The figurative mushrooming of methods available to quantitative ecologists who are interested in demographics has led to some incredible advances in biometrics (the actual statistical models and methods) and our understanding of complex systems.

However, the advance in models for parameter estimation causes two problems. First—*a user problem*—most of the new models are very complex and have many more parameters to estimate than their predecessors. Our experience cautions that the novice users of estimation models should start with simple models and proceed to complex models if it appears data will allow. And, second—*an instructor problem*—the array of potential models causes problems for instructors of parameter estimation courses—there is literally no way to cover all the models during a one semester course. So, we have chosen a set of model structures for mark-recapture and survey scenarios that we think form the basis for all subsequent analyses—if you understand the methods included in this primer, you should be able to read a bit more information and work up to the more complex methods available.

As you move forward from this primer, we will caution you—just because something is more complex does not mean it is better. In fact, that idea will become very important as we discuss model selection in a statistical sense!

We should also note that your authors are visual learners. For that reason, we provide a wealth of diagrams, flow charts, and other images that we believe will help many students. In our experience, students in wildlife biology, conservation biology, and resource management *tend* to be visual learners—that is, the majority of students in these fields of study (relative to engineers or mathematicians for example) tend to learn better when looking at images rather than text. So, for the visual learners in the world, we provide visual stimulation. If you are a verbal learner who learns more efficiently from reading and pondering text, you may focus on the textual explanations. But, perhaps the imagery will also be of use when interacting with colleagues who are visual learners.

We hope you find this primer useful as a first step towards a basic understanding of estimation of parameters for fisheries and wildlife populations.

Our reviewers

We are very appreciative of comments and critiques provided by a team of reviewers who evaluated chapters for us. Our book is better for their suggestions.

Mary Bomberger Brown, PhD
Christopher Chizinski, PhD
Wanlop Chutipong
Rajendra Dhungana
Martin Hammell, PhD
Lucia Coral Hurtado
Jocelyn Olney Harrison
Greg Irving
Joel Jorgensen
Kathryn McCollum
Dusit Ngoprasert, PhD
Mark Pegg, PhD
Andrew J. Pierce
Chanratana Pin

Erin Roche, PhD
Stephen Siddons
Maggi Sliwinski
Jennifer Smith, PhD
Andrei Snyman
Jonathan Spurgeon
Kirk Stephenson
Niti Sukumal
Christopher Salema
Saranphat Suwanrat, PhD
Supatcharee Tanasarnpaiboon
Mark Vrtiska, PhD
Cara Whalen

Inspiration from our colleagues

We based this book on lectures that we have given in graduate and undergraduate courses. As teachers refine lectures over the years, we often do not record the source of the interesting examples that we find in on-line lecture postings from our colleagues who teach similar courses. We have tried our best to locate the original sources of drawings, figures, charts, examples, and unique descriptions that we've collected over the years. If we have missed an attribution, please let us know so that we can make the correction in a future version of this book.

Many of the teaching materials (copies of handouts from a 2002 workshop, an on-line PDF that no longer is posted on-line, lecture materials shared via email) that we consulted as we wrote this book were not available in a citable source. Further, because instructors don't use citations frequently in their lecture materials, it is very difficult to trace the original source of materials (we noticed some highly correlative connections in many instructors' on-line materials that appear to be based on original teaching examples of Gary White, for example—but who can tell for sure). So, we chose to simply acknowledge the colleagues who we know or believe to be the source of many of the fine ideas that we have modified, expanded upon, or refined in our primer. Thus, each chapter begins with a footnote to document the influence that our fellow quantitative ecologists have had on our teaching and on the ideas presented here. We are grateful for those colleagues who take time to think about how to present material—especially those who discover ways to make highly quantitative ideas more palatable to the beginner.

For more information on topics in this book

Amstrup, S. C., T. L. McDonald, and B. F. J. Manly. 2005. Handbook of capture-recapture analysis. Princeton Univ. Press: Princeton, NJ.

Anderson, D. R., 2008. Model Based Inference in the Life Sciences: a primer on evidence. Springer: New York, NY.

Buckland, S. T., et al. 2001. Introduction to Distance Sampling: estimating abundance of biological populations. Oxford University Press.

Conroy, M. J., and J. P. Carroll. 2009. Quantitative Conservation of Vertebrates. Wiley-Blackwell: Sussex, UK.

Cooch, E., and G. White. 2014. Program MARK: a gentle introduction, 12th edition. *Online: http://www.phidot.org/software/mark/docs/book/*

Donovan, T. M. and M. Alldredge. 2007. Exercises in estimating and monitoring abundance. *Online: http://www.uvm.edu/rsenr/vtcfwru/spreadsheets/abundance/abundance.htm*

Donovan, T. M. and J. Hines. 2007. Exercises in occupancy modeling and estimation. *Online: http://www.uvm.edu/rsenr/vtcfwru/spreadsheets/occupancy/occupancy.htm*

Krebs, C. J. 1999. Ecological methodology, 2nd edition. Benjamin/Cummings, Menlo Park, CA.

Millspaugh, J., and J. M Marzloff. 2001. Radio tracking and animal populations. Academic Press.

Thompson, W. L., G. C. White, and C. Gowan. 1998. Monitoring vertebrate populations. Academic Press, San Diego.

Thomson, D. L., E. G. Cooch, and M. J. Conroy. 2009. Modeling demographic processes in marked populations. Springer Press: New York, NY.

Williams, B. K., J. D. Nichols, and M. J. Conroy. 2002. Analysis and management of animal populations. Academic Press, San Diego.

Foreword

Abundance estimation of animals is considered a fundamental tenet for both empirical and applied research. Critical answers relating to topics from evolution of animal behavior to species conservation status are based at least partially on knowing how many individuals there are in an area or population. Ironically, despite decades of realization of the need for reliable abundance estimation we have seen only recently (last 20 years) an explosion of research assessing the relative value and theory of estimation techniques, assumptions, and mathematics. The old adage of "keep it simple stupid," becomes a bit more like "ignorance is bliss." Fundamental techniques, such as mark-recapture and related techniques, which have been around for a very long time, now have received much needed evaluation and updating. As a result, the underlying math required to develop estimations has become far more complex and difficult for many field biologists to master.

Over the years a number of important treatises on these techniques and their application have been released. Most notable is what many of us consider the "bible of abundance estimation," Williams et al. (2002). For those of us involved in training future conservation scientists at universities we consider that book is fundamental, but mathematically beyond almost all our undergraduates and many graduate students creating a need for a bit more user-friendly versions. Several subsequent volumes directed at various levels of experience and expertise have been released, including one I authored with Mike Conroy in 2011. However, there has been still a gap in getting students or field biologist's in the door of understanding the basics and philosophy of abundance estimation.

The authors of this Primer have set out to create a way into the world of animal abundance estimation for those who are not mathematicians, but are instead field biologists who need to properly apply techniques and analyses of their data. I have known Larkin and George for many years and both are highly dedicated field researchers, but more importantly in undertaking this endeavor they are showing their true colors in being even better teachers.

The best techniques and analyses in the world are of no use if those who need them the most lack the resources to apply them in fundamentally sound ways. By providing a simple, clear, and concise description of the foundations of mark-recapture theory and application, Larkin and George provide a place to start and important tools to aspiring biologists as they apply abundance estimation to their particular research problems. In the tradition of education first, Larkin and George are providing their intellectual property to those who need it for free. I recommend that those of you who use this book thank them by providing feedback to them to help make this a living rather than static product.

John P. Carroll, University of Nebraska
June, 2015

Ch. 1

Models of populations[1]

"Modeling is the best because you have to look hot, which comes easy to me, you know. I'm blessed with that."
-- Ashton Kutcher

Questions to ponder:
- *How might we use models to better understand a population?*
- *Can I use my parameter estimates in a population model?*
- *How complex should I make my model?*
- *What is the difference between an analytical model and a statistical model and a conceptual model?*
- *What is a sensitivity analysis?*

Me? A modeler? *Never!*

You are here, which suggests that you must be interested in the study of wild animal populations. The rest of this primer is dedicated to descriptions of how to estimate parameters for populations. But, in life it is sometimes good to take a step back, and ask, *"Why am I doing this?"*

Why would you want to estimate density, or population size, or probability of survival, or the probability of moving from A to B? Undoubtedly, there are a zillion reasons for studying populations, but all of those reasons can be boiled down to one statement—and it explains why we all became population ecologists: **we want to understand our study population.**

Is our population growing? Why? Is our population going extinct? Why? Is a segment of our population exhibiting some interesting behaviors or dynamics? Why?

A **model** is a simplification of the real world that allows study of the real world. A well-known statement about modeling suggests that "…all models are wrong, and some are useful (Box and Draper 1987)." How do we construct a useful model? Should we?

---

[1] *With thanks for content to Anthony Starfield, Therese Donovan, and Michael Conroy*

Figure 1.1: *Ashton Kucher, a model, demonstrates the use of a Nikon camera.*

Modeling

We want to convince you that you should start to examine your population through modeling. No, not the type of modeling that Ashton Kucher is talking about! Ashton spends his time as an actor, and has modeled clothes and cars for TV commercials. That is a different type of modeling. But, even Ashton Kucher's modeling does the same thing that a population model can do—*it can show us something.* Ashton shows us what we might look like if we purchased a Nikon camera—if we were as handsome as Ashton (Figure 1.1)! And, a population model can show us insights to our wildlife population to support decisions for management.

You may have never thought about using models in your biological work, because they may seem complicated and scary. But, models can be simple and fun—if we approach them correctly. As we start to think about using models in populations, let us ask you a question (Starfield 2005):

Look around the room in which you are sitting. How many balls could fit into the room?

Seriously, figure it out. We'll wait!

Figure 1.2: *A room and a ball.*

You now have an estimate of how many balls will fit in your room. Look back at your notes that you used to solve this question—and realize that *you are looking at your model!*

Certainly, we all came up with a different answer with regard to how many balls fit into our room. Of course, we are all sitting in different-sized rooms. Your room might be small, like an office, or might be large—like a large classroom. And, what kind and size of ball did you imagine? If you thought about a ping pong ball, your answer will be different from someone who imagined a basketball.

So, the basic structures that we used for our model varied—small or large rooms, and small or large balls. Now, how did you stack the balls in your conceptual model? Did you estimate the diameter of your ball and figure out how many balls would fit along the floor, wall-to-wall? Then, did you just multiply that by how many balls could be stacked on each other to reach the ceiling?

MODEL STRUCTURE

What kind of ball were you supposed to fit into your room? Your choice may have depended on where you live. Did you select one of these types of balls?

France:	tennis ball
England:	cricket ball
Thailand:	takraw ball
China:	soccer ball
Mexico:	*Pelota mixteca* ball
United States:	baseball
India:	throwball
Scotland:	golf ball
South Africa:	rugby ball
Australia:	beach ball
Hogwarts:	quaffle
Italy:	bocce

Another, more complicated, way to stack balls is to realize that if you have a first level of balls, that the next level will actually fit down into the 'pockets' of space created by the round balls in the first level (Figure 1.3). So, you can actually stack more balls in the room if you account for this, as you can see in the figure.

Did you account for the desks and bookshelves in your rooms? Or, did you pretend your room was vacant—just walls, floors, and ceilings? We can simplify reality or try to incorporate details in our models—and we will continue to see this theme throughout the book.

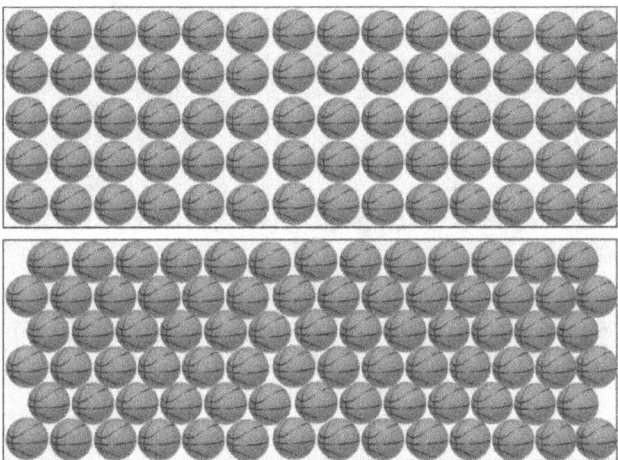

Figure 1.3: *Two different methods (top and bottom) for stacking balls in a conceptual model. The pattern affects the model estimate for the number of balls that can fit in a room.*

An important question: "why did you construct your model in the way that you did?"
Why didn't you account for all of the chairs and the lights and the desks in your room, for example? Your answer will most likely be: "I thought this was just a simple exercise…I don't have all day to spend on this chapter!"

And, you've just expressed the purpose for your model—you wanted to get a rough estimate of how many balls fit in the room. You weren't worried about being incorrect.

What if you were an engineering firm and you needed to know exactly how many balls could fit in the room, because you needed to budget for balls in your annual budget? You would spend more time on this exercise, and your model would be more detailed, correct?

Regardless of how we approach them, models are—at their essence—<u>tools</u> to help us accomplish something.

Real World vs. Model World

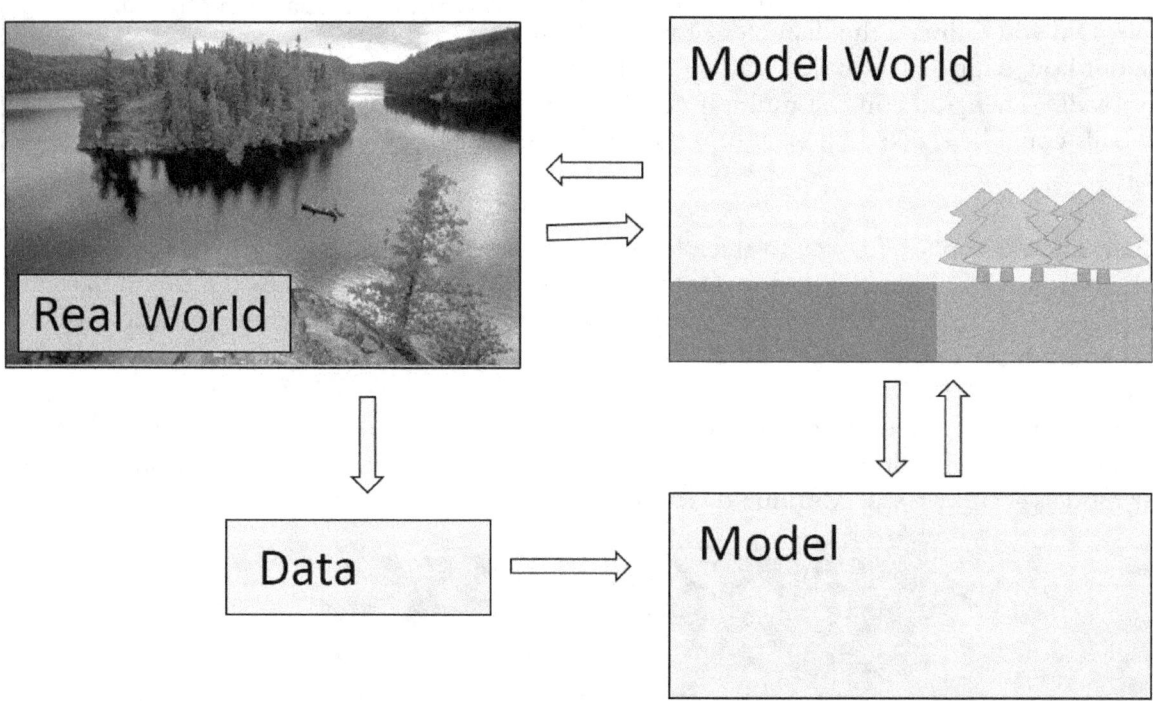

Figure 1.4: *The relationship between the Real World, the Model World, a Model, and Data. Figure after Donovan and Starfield (2015).*

Professor Tony Starfield, from the University of Minnesota, explains the modeling process using the framework shown in Figure 1.4 (Donovan and Starfield 2015).

The Real World is bumpy, detailed, curvy, dirty, complex, and full of information—too much information, really. Even a photograph is a simplification of the Real World! A modeler creates a Model World to represent the real world, for the purposes needed by the current research question. The Model World exists in the mind of the modeler—or it may exist in a series of sketches or flow charts that describe the model world as the modeler "sees" it. The Model World is a simplification of the Real World—some bits are missing completely. For example, in our diagram of the Model World (Figure 1.4), you don't see any fish or any islands that are in the Real World. Others, such as the trees and the lake in this diagram, remain in the Model World.

The next step is for the modeler to place the Model World into the Model. To create a model, the modeler will use an **algorithm**, a process or formula used to solve a problem. If you are using a computer for your modeling process, the algorithm is how you place the Model World in the computer. The algorithm could be a simple equation, or a complex set of equations, or it could be a simulation—our model can take many forms.

For example, if the purpose of our model was to predict the population size (N) of pine trees in this area, we could use this simple model (where N is the population size in time t and r is the intrinsic rate of growth for the population):

$$N_{t+1}=N_t * r$$

But, where do we get values for N and r to use in our model? We have to return to the Real World (Figure 1.4). We would sample the trees to estimate the population size, and we would need to investigate the population to determine birth rates (b) and death rates (d) to get a value for r, because we know that $r = b - d$.

So, the Real World informs us as we consider the Model World. And, the Real World provides Data that we use in our model. We use the Model World to create our Model. The Model then informs us about the Model World. It is important to note that the model does not directly inform us about the Real World—we make inferences about the Real World from our modeling process, but we have to remember that we created our Model by thinking about the simplified Model World—not the real world. But, of course, we hope that our model is useful as we make decisions about how to manage populations in the Real World.

Types of models

There are many types of models that we may use in our career as an ecologist. First, we can imagine that the Model World can be represented in two ways:

> **Conceptual** – a representation of the real world in our brains, or a collection of concepts that we believe to be important. At the conceptual stage the model exists in verbal descriptions or thoughts—literally in our imagination.
>
> **Physical** – abstractions drawn on paper or built in three-dimensions (a 'model boat').

A wise, rewarding first step toward the design of a new research project is to place an image of your study species in the middle of a sheet of paper. Then, place words or images of important factors that affect reproduction or survival of the species. Perhaps "habitat", "predators", "rainfall", "food", and other words might appear. Now, draw linkages to depict the connections that exist in the Real World. For example, rainfall might affect food availability, which impacts both survival and reproduction of your species. Habitat provides cover, which affects survival.

When you are done, you have your first physical depiction of the Model World of your species in its ecosystem.

Next, we can take conceptual or physical Model Worlds and turn them into models:

Graphical – a type of physical model that shows relationships (weight of an elephant during its first 5 years of life for example). Even without a specific equation, we can look at the figure and predict the weight of the elephant after 10 years, assuming growth continues at the same rate.

Analytical or numerical – mathematic equations (can be coded on a spreadsheet or other computer program) that provide a numerical outcome, or result. We might, for example, be able to derive a formula for the elephant's change in weight (in kg) over time (where x = age in years, Figure 1.5):

Figure 1.5: *A young African elephant (Loxodonta africana) drinks with adults at a water hole in Namibia. Photo by Kelly Powell, used with permission.*

$$\text{Weight} = 100 + 85x$$

Statistical – a model that uses data to estimate the value of a parameter (e.g., survival probability, mean weight, etc.). For example, we can develop a statistical model to estimate annual survival rate, S, for elephants using the number tagged (n_{tagged}) and the number of tagged elephants that survive the year (n_{alive}):

$$S = \frac{n_{alive}}{n_{tagged}}$$

That simple fraction is a statistical model to estimate S. The formula for a mean of a set of body weights, or mass ($mass_i$), for a sample of n individuals is also a statistical model:

$$\bar{x} = \frac{\sum_{i=1}^{n} mass_i}{n}$$

Reasons for using modeling

The reasons we might want to use a model are abundant, but here are some examples of problems that are commonly addressed with a model (Starfield 1997):

1—Our resources are scarce. In some cases, it is easier or cheaper to conduct a simulation of a population rather than run an expensive experiment. Or, we might use a model to conduct an initial analysis of a population so that we can determine which vital rate dynamics are the most important. Based on that simulation, we might, for example, decide to spend our money and time in the field (in the Real World!) collecting data about reproduction rather than adult survival.

2—We need support for a management decision. Which scenario is more likely? What will be the outcome if we modify habitat or conduct a translocation to a population?

3—We need to 'visualize' a complex situation and learn ecology! Sometimes the process of modeling may be more important than the end-result—that is, we might learn about a population by sitting down and thinking about what parts of the Real World are important to include in the Model World.

4—We want to predict the future. Many models are predictive—we use current information to project forward to predict the population size in 50 years, for example. Of course, we must remember that all predictions are exactly that: *predictions*!

The modeling process

Take a few minutes and think about your study population—how might you use a model? How would the modeling process help you learn about your population?

If we have convinced you that a modeling exercise for your population could be useful, there are seven basic steps to follow (Starfield 1997):

1—Define the problem. What is the purpose of the modeling exercise?

2—Identify important variables. This is where we construct our Model World. And, this is a good time to start a simple diagram of your population with various pieces of information that you think are important to consider—to help you answer your question.

3—Create the model by developing a formula that describes the output that you desire (future population size, for example).

4—Solve the model. Run the simulation, or calculate the value of the output described by your equation. For now, you can use estimates available or "best guesses" of the values for the variables in your model.

5—Interpret the results you receive. Do they make sense? If not, is your model constructed properly? What does this mean for your population?

6—Investigate your model with a sensitivity analysis. How sensitive are your results to small changes in the values of parameters? Play with your model! Increase the survival rate slightly—what happens to population size? Now, decrease survival, and see what happens.

7—Conduct a thought-analysis of assumptions. Think about your Model World and the assumptions you've made as you simplified the Real World. As you make inferences about the Real World from your model results, these assumptions are important to remember, and you must share them with your colleagues as you report your model results.

Problem context is important

Should you model? If so, how?

Although we've pushed you to consider creating a model for your population, there are situations for which you don't need to model...or for which you should not model. The answer to the question, *"Should I model?"* depends on two things: how much data is available for your system, and how much you understand the system (Figure 1.6).

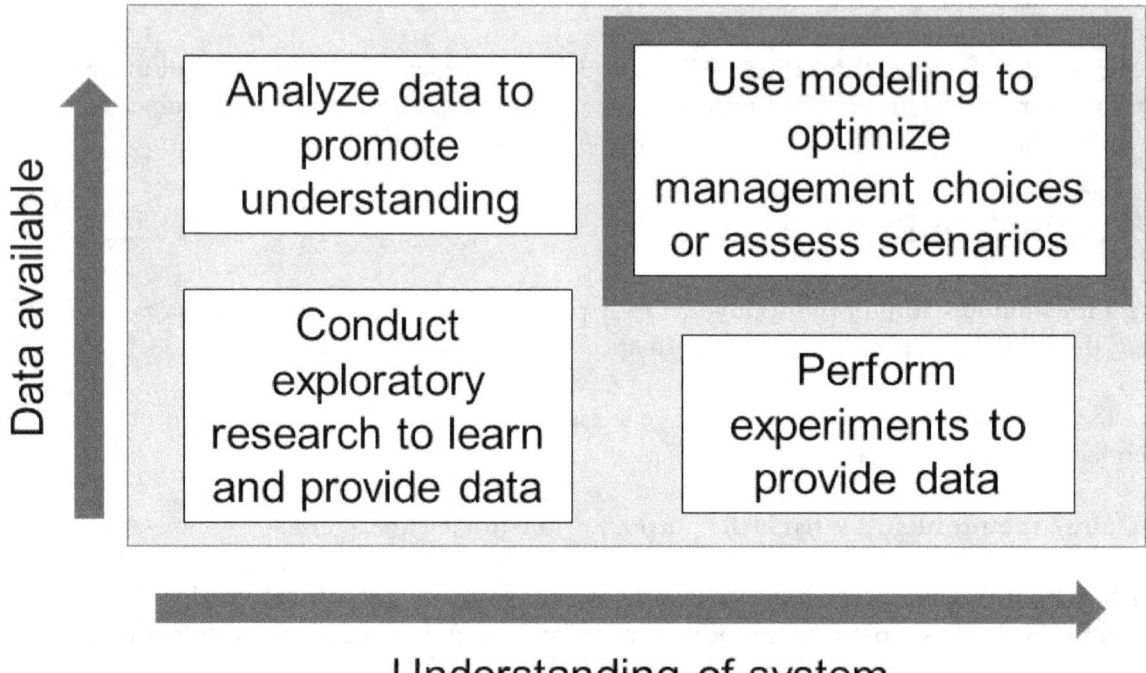

Figure 1.6: *Depiction of possible states related to the availability of data and our understanding of a biological system. Our goal, or ideal situation, is the darker area at top right where we have good knowledge of our system and a lot of data available to parameterize a model. However, models can be used for different purposes along the two-dimensional gradient. Based on concepts in Starfield (1997).*

Of course, we'd like to understand a system completely, and we'd love to have a lot of data available. In fact, if we already understand our system, it is probably because we have access to a large amount of data. If you have little data and no understanding of your system (lower left of the graphic), you can't expect to properly model the processes that drive the dynamics of your population. In that circumstance, you would need to start with an exploratory, conceptual model to help you learn more about the dynamics of your system. You should also go collect more data, and a model might help you understand which data you should collect first.

If you have a large amount of data collected, but don't understand the system, a model can help you combine the various types of data to understand your system better.

If you are in the often enviable position of having a lot of data and a lot of understanding, you may be in the position of being able to make some defendable management decisions—a model could help you decide which management option would give you the optimal benefit. How can you best meet your objectives? Should you provide nesting habitat to increase productivity, or should you focus on alternatives that improve survival rates of adults?

Conclusion

Modeling does not have to be a daunting process. In fact, every ecologist should be able to describe a conceptual model of their system. Conceptual models—the Model World—are simplifications of the Real World, and conceptual models inform the construction of the model that we use to provide information about our population. In many situations, the process of building a model can be more important, or at least as important, as the results that we obtain from our model. To create a model, we must think critically about our study population and its environment—what is important to include in the Model World?

As you move forward, you will encounter many, many types of population models—some describe what happens to populations under scenarios of competition or predation, for example. You will find examples of simple population models in any ecological textbook. But, we encourage you to build your own model to describe *your* study population.

The fundamental model that we will return to throughout this primer (see Chapter 2) is the "BIDE" model of population growth which describes population growth (ΔN) as a function of Births, Deaths, Immigration, and Emigration:

$$\Delta N = B - D + I - E$$

Most populations are based on the simple idea that only these four dynamics cause changes to population size over time. As we build our models, we may modify the fundamental model to suit our circumstances and our population of interest. To the point of this primer, we can see that we have a very basic need if we hope to determine how our population changes over time—we must obtain estimates of density (to learn about population size, N), survival (to learn about death rate), and movement (to learn about immigration and emigration). *So, let's get started!*

References

Box, G. E. P., and N. R. Draper. 1987. Empirical model building and response surfaces. John Wiley & Sons, New York, NY.

Starfield, A. M. 1997. A pragmatic approach to modeling for wildlife management. The Journal of Wildlife Management 61: 261-270.

Starfield, A. M. 2005. Principles of Modeling: Real World - Model World. *In* Starfield, A., and T. Donovan. Principles of modeling with spreadsheets. On-line: http://www.uvm.edu/rsenr/vtcfwru/spreadsheets/?Page=pom/pom.htm

Starfield, A., and T. Donovan. 2015. Principles of modeling with spreadsheets. On-line: http://www.uvm.edu/rsenr/vtcfwru/spreadsheets/?Page=pom/pom.htm

For more information on topics in this chapter

Conroy, M. J., and J. P. Carroll. 2009. Quantitative Conservation of Vertebrates. Wiley-Blackwell: Sussex, UK.

Donovan, T. M. and C. Welden. 2002. Spreadsheet exercises in ecology and evolution. Sinauer Associates, Inc. Sunderland, MA, USA. On-line: http://www.uvm.edu/rsenr/vtcfwru/spreadsheets/?Page=ecologyevolution/ecology_evolution.htm

Nicolson, C. R., A. M. Starfield, G. P. Kofinas, and J. A. Kruse. 2002. Ten heuristics for interdisciplinary modeling projects. Ecosystems 5: 376-384.

Shenk, T. M., and A. B. Franklin, eds. 2001. Modeling in natural resource management: development, interpretation, and application. Island Press.

Williams, B. K., J. D. Nichols, and M. J. Conroy. 2002. Analysis and management of animal populations. Academic Press, San Diego.

Citing this primer

Powell, L. A., and G. A. Gale. 2015. Estimation of Parameters for Animal Populations: a primer for the rest of us. Caught Napping Publications: Lincoln, NE.

Biologists use marks in the marginal scutes as individual identification of western painted turtles (Chrysemys picta bellii) in western Nebraska, USA. Photo by L. Powell.

Ch. 2

Context and inference[1]

"Thinking is the hardest work there is, which is probably the reason why so few engage in it."
  -- Henry Ford

Questions to ponder:
- *What is the difference between induction and deduction?*
- *What is the difference between a biological and a statistical population?*
- *What is binomial sampling?*
- *How can I help ensure that my sample is representative of my study population?*

Some context

The population. The context of everything we will cover in this primer is the wild animal population. We know from basic knowledge of population dynamics that the growth of a population (ΔN) during a given time interval (Figure 2.1) is equal to the number of births (B) and immigrants (I, the additions to the population) minus the number of deaths (D) and emigrants (E, the subtractions from the population).

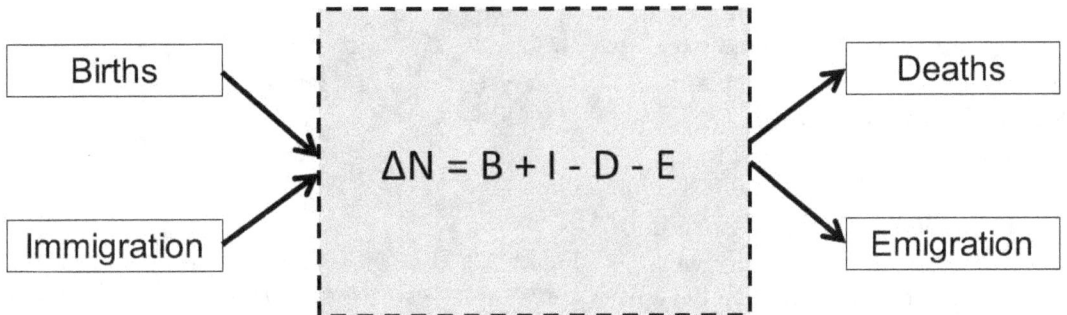

Figure 2.1: *The basic "BIDE" model of population dynamics in which the change in population size (ΔN) in a given time period is equal to the number of births (B) and immigrants (I), minus the number of deaths (D) and emigrants (E). Here, the dotted box represents an open population.*

---

[1] *With thanks for content to Michael Conroy, Gary White, and Mary Bomberger Brown*

So, what is important to know about this population if you are a population biologist?

We need to know the size of the population at a given time, and we need some estimation method to accomplish this goal. Then, we need to know something about the per capita birth rate of the population. We'd like to know something about movements in the population and we'd like to know something about survival and mortality in the population. Admittedly, there are few populations for which we know all of these things simultaneously, but such an achievement can certainly be our goal.

How do we estimate N? How do we determine the productivity (usually defined as the number of offspring produced per adult) of the population? How do we identify movement probabilities? And, how do we estimate the probability of survival for the period in which we have interest?

The answer is that we can use a variety of tools—surveys, mark-recapture, or radio-telemetry methods can all provide information from which we can derive estimates of these parameters. And, that is exactly what this book is about—how do we use these tools, and how do we decide which tool to use?

The management cycle. Most parameter estimation occurs in the context of providing information for management decisions. Is the population large enough to sustain a harvest event? Should we spend money to renovate habitat for this species to increase birth rates?

Our goal is to make decisions for management that are based on reliable information, and we should make better decisions when we have a better understanding of a biological system. A simple description of a management process is that we encounter a problem or a question. We gather data to inform the decision to be made with regard to this management problem. We assess the information and make the decision, and we put that decision into action on the ground—e.g., we get out the torches to conduct a prescribed burn, or we plant food plots, or we thin the forest, or…whatever we believe we need to do.

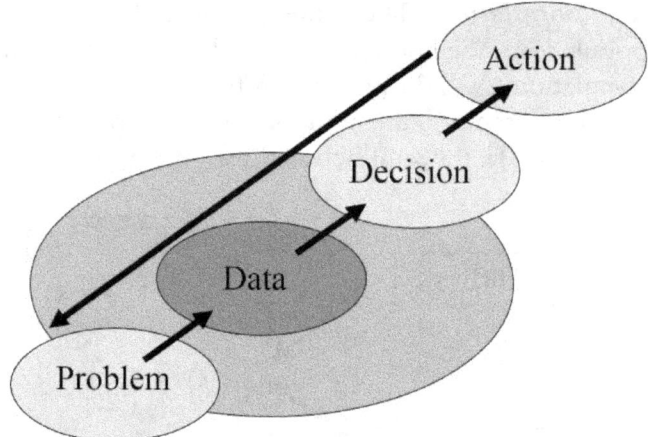

Figure 2.2: *A cursory view of the management cycle in the context of collection and analysis of data to inform a decision. Emphasis shown on the Data step to reflect the goals for this primer.*

This primer is aimed at the 'data' portion of the management process. How do we gather and analyze the data that we have to provide reliable information for decision-making and eventual action?

If we make our diagram of the management process a bit more detailed, we might draw something that matches the cycle described by adaptive resource management (Figure 2.3). Assess a problem—design a management approach—implement the management approach—monitor a critical feature that will be impacted by management—evaluate the results of our management—and adjust our management approach based on what we learned from the process thus far. Is there still a problem to solve or not?

It is useful to note that this cycle of adaptive resource management (Figure 2.3) is very similar to the scientific method. We are familiar with this basic cycle: we make observations, develop hypotheses to explain our observations, make predictions, and perform an experiment. After gathering data, we analyze and determine how the data match our predictions. We reconsider, if needed, and move forward.

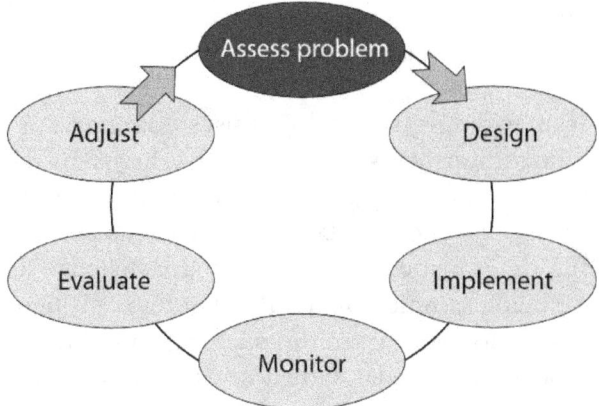

Figure 2.3: *Representation of six steps in the cycle of adaptive management.*

Four critical elements of the scientific method and management cycle:

- <u>Characterizations:</u> we make observations, we define problems, and we take measurements.
- <u>Hypotheses:</u> we suggest theoretical explanations of observations and measurements
- <u>Predictions:</u> we deduce, from our hypotheses, reasonable expectations from an experiment
- <u>Experiments:</u> we test predictions and our hypotheses

Now that we have thought about the context of our research, we need to consider the inference of our results. In the rest of this chapter, we will cover 4 basic topics before we launch into parameter estimation:
- Random and representative samples
- Types of inference
- Precision and bias
- Binomial coefficients

Populations and samples

We may define wildlife populations in many ways: spatially, temporally, taxonomically, and demographically. A biological population is often defined as a group of organisms in a certain place, or space, at a certain time.

However, a statistical population is a bit different, and it is important to keep track of whether we are talking about a statistical population or a wildlife population.

A **statistical population** can be defined by dividing up time and space into elements that can be independently sampled---we call these elements 'sampling units.' Thus, if our research design or sampling method (i.e., size and type of net, trap type, or survey type) does not allow sampling of the entire biological population, inference is NOT about the entire biological population. The inference is to the statistical population that we sampled.

As an example, if a biologist uses floating traps to catch turtles in a pond for a mark-recapture estimation process, she knows that her traps do not capture the small turtles well. In fact, pond turtles do not get up on the floating traps until they are large enough to have less worry about predators. The population of turtles, as a **biological population**, includes the small turtles. But, the statistical population in this instance does not include the small turtles, because they are not being sampled by the floating traps. When the biologist obtains her estimate of the 'population size' from the mark-recapture method, her estimate has inference to the statistical population—the large turtles. She knows nothing about the population size of the small turtles.

So, samples are important to the processes that we will cover for surveys or mark-recapture-type analyses. Our problem, as wildlife biologists, is that we normally cannot collect, measure, and observe the entire population. We are limited by logistics. The solution is that we can take a sample and we can make inferences about the population from the sample. But, this is possible only if our sample is representative of the population.

Again, we can define our statistical, **study population** as the group of animals to which inferences will be made. The **sample** is the subset of the population for which data is available or collected.

We can spend a lot of time thinking about our study design—is our sample representative of the population? How can we use random selection of individuals to assure that our sample is not biased in some way?

The mismatch of a non-representative sample and a population is shown by a famous story about a group of blind men and an elephant. The ancient story was set as a poem, The Blind Men and the Elephant (Figure 2.4), by the American poet John Godfrey Sax (1816-1887).

Figure 2.4: "*Blind monks examining an elephant,*" *an ukiyo-e print by Hanabusa Itchō (1652–1724). Available as public domain image from the Library of Congress (LC-USZC4-8725).*

A portion of Sax's poem reads:

> *It was six men of Indostan*
> *To learning much inclined,*
> *Who went to see the Elephant*
> *(Though all of them were blind),*
> *That each by observation*
> *Might satisfy his mind …*
>
> *And so these men of Indostan*
> *Disputed loud and long,*
> *Each in his own opinion,*
> *Exceeding stiff and strong,*
> *Though each was partly in the right,*
> *And all were in the wrong!*
>
> <u>Moral:</u>
> *So oft in theologic wars,*
> *The disputants, I ween,*
> *Rail on in utter ignorance*
> *Of what each other mean,*
> *And prate about an Elephant*
> *Not one of them has seen!*

Inference

A dictionary definition of 'inference' is: "the act or process of reaching a conclusion about something from known facts or evidence." There are two types of processes used in science to draw inferences. We may refer to these as two types of reasoning.

First, **deduction**, or **deductive reasoning**. When we use deduction, we are reasoning from the general to the particular. This type of reasoning is central to the study of logic, and deductive inferences must be correct.

For example, we can state a truth, or we can posit that all birds have feathers. We can also state a second truth, or second position, that a duck is a bird. If these two things are true, then we can infer that a duck has feathers. It must follow as a truth, because we have stated that all birds have feathers, and we have stated that a duck is a bird. Thus, ducks have feathers. The only way to falsify this deduction is to show that one of the posits is, in fact, false.

A second type of inference is called **induction**, or **inductive reasoning**. When we use induction, we reason from the sample to the population…or from the specific to the general—the opposite of deduction. Induction may also be used to reason from the past to the future: a weather forecaster uses induction to make predictions in a weather forecast.

Induction is central to the study of statistical inference, and hence the rest of this primer. When we use induction, we allow for the possibility that our conclusion may be false.

For example, we might observe that a population has grown at a rate of 4% for the past 10 years. Therefore, we might use induction to predict that the population will continue to grow at 4% per year in the future.

Or, if 95% of a sample of fish has a snout length of 8-10 cm, we may infer that 95% of the entire population has a snout length in the range of 8-10 cm. That is, we use induction to assess the population by assessing our sample. If our sample is indeed representative of the population, our induction should be correct.

Accuracy, bias and precision

Gary White, professor emeritus at Colorado State University, has noted that "Statistical estimation is like shooting a single arrow at an invisible target and inferring the location of the bull's eye based on where the arrow lands."

So, a good estimator is like a good archer—a good estimator is accurate, but as we will see the definition of accuracy is not as straightforward as we might imagine.

We can distinguish between three concepts that are sometimes confused (White et al. 1982): accuracy, bias, and precision.

A dictionary definition of **accuracy** is "the condition or quality of being true, correct, or exact; freedom from error or defect." When we sample nature and use inductive inference, it is impossible for us to know truth. As White et al. (1982) noted, an archer can approach the target and see the distance between truth (the bullseye) and the estimate (the grouping of arrows). A biologist, however, is never able to access the true value of a parameter for a population. Instead, we rely on two measures to help us assess our estimation: *bias* and *precision* (Figure 2.5).

> **Definitions:**
>
> *Parameter:* (θ, *theta*), a random variable; an unknown quantity or constant characterizing a population (e.g., density, population size, capture rate, survival)
>
> *Estimate:* ($\hat{\theta}$), a numerical approximation of a true population parameter (e.g., density estimate, $\hat{D}$)
>
> *Estimator:* a mathematical formula used to compute an estimate

A **precise** estimate is one that has very little uncertainty. If we obtain a precise estimate, we would predict that a second, third, and fourth sampling effort would return a very similar estimate—that is, our results would be very repeatable (Figure 2.5). Statistically, that might mean an estimate with a small standard error, or a narrow 95% confidence interval surrounding the estimate.

Bias, on the other hand, describes the match between the estimate and the true value of the parameter (Figure 2.5). Of course, we want our estimate for a parameter to be free of bias. Bias can occur from two sources: *small-sample bias* and *model bias*. We will discuss in Chapter 8, for example, how the simple Lincoln-Petersen estimator for population size may be biased at low sample sizes. And, throughout this primer we will refer to assumptions of estimators that relate to study design issues that may create opportunity for model bias. For example, if the marking method affects the survival of an individual, the estimate of survival for the population will be biased. Or, if animals become trap-shy after encountering a trapping event, the estimates of population size may be biased if capture probabilities are not structured appropriately in the model.

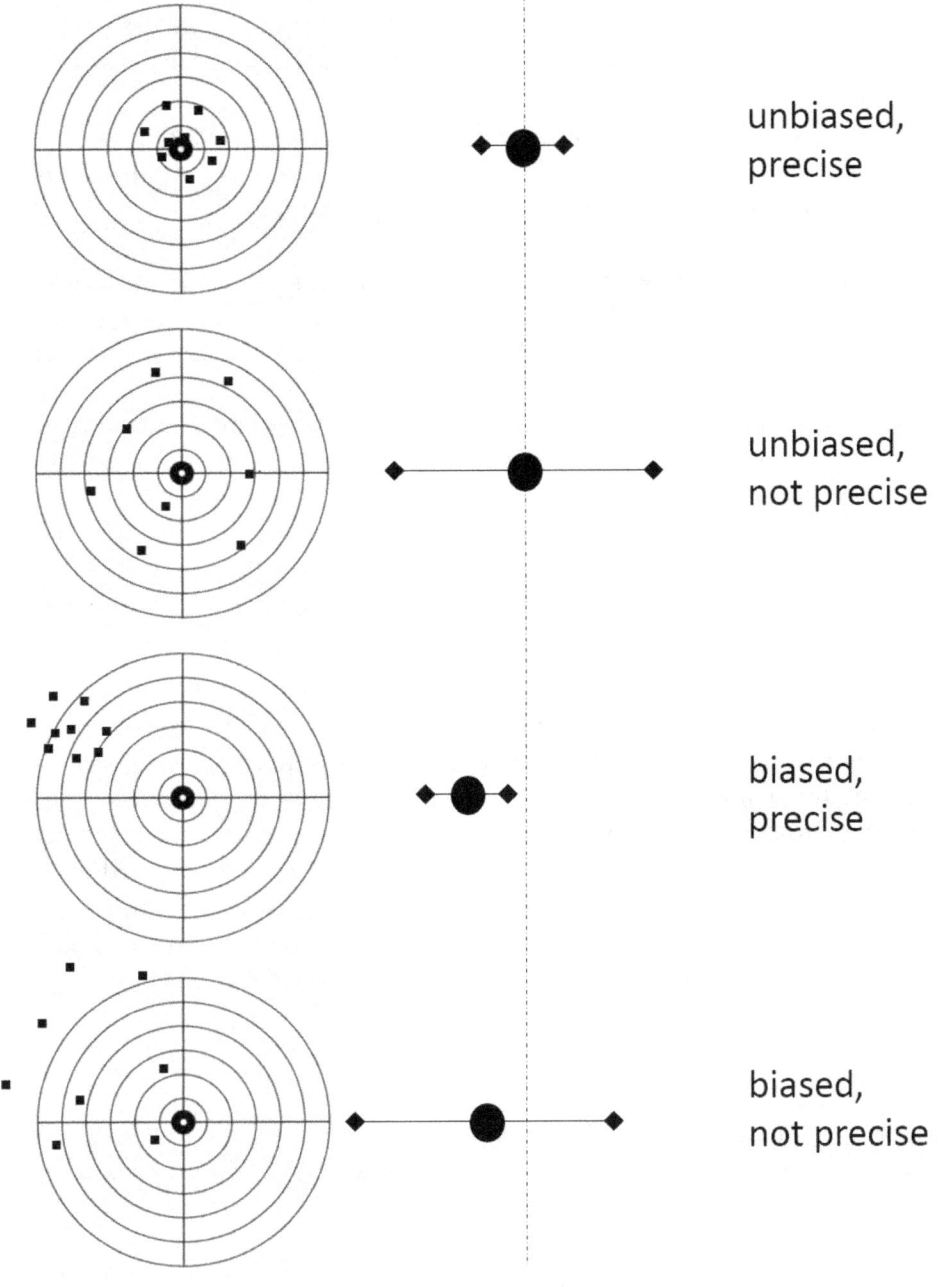

Figure 2.5: *Illustrations of the concepts of bias and precision with shot patterns (left) (bull's-eye is truth; after White et al. 1982 and Williams et al. 2002) and 95% confidence intervals (right) (dotted line shows true parameter value, circle shows parameter estimate). The typical goal of an archer and an ecologist is shown at top as the combination of a lack of bias and high precision.*

Generally, we might state that our goal, when estimating parameters, is to be **accurate**. Although "accuracy" conveys the general sense of our goal, *quantitative ecologists and biometricians do not agree on the definition of this term*. And, it is important to understand that "accuracy" may mean different things to different people. To some, accuracy is the same as the lack of bias (Zar 1999, Walther and Moore 2005)—*the distance of the estimate from the true parameter value*. Under this view, the top two scenarios in the bull's-eye figure are *accurate*—the estimate is the same as the true value, regardless of the level of precision. Others (White et al. 1982, Williams et al. 2010) describe accuracy as *the blending of bias and precision*. With this definition, only the top scenario in the bull's-eye figure is *accurate*, because it is not biased and it is precise. The introduction of *either* process, bias or lack of precision, makes the other scenarios inaccurate under this view.

What to do? We suggest that readers solve this dilemma by using a direct approach—refer explicitly to either bias or precision, or both, when describing estimates. **Avoid the use of the term "accuracy" because of the confusion surrounding its meaning.**

In the 'game' of estimation of parameters for wildlife populations, we value both precision and the lack of bias. It may be difficult to always have unbiased and precise estimates, however. So, it can be generally said that we value the lack of bias over precision—although both are certainly desired. Precision can typically be improved through improvements in sampling, when possible. But, a biased estimate cannot be improved without reevaluating the entire sampling and analytical scheme. A focus of this primer will be to help the biologist reduce the bias and increase the precision of demographic estimates with a focus on sampling effort, study design, and use of proper estimators.

Conclusion

Parameter estimation is not something we do in isolation—we estimate the value of parameters of wildlife populations to learn about their ecology. And, we often estimate parameters as part of a management program. As such, we must view parameter estimation as a part of the scientific method and the management cycle. Our estimates are derived from samples of marked or observed animals, and our goal is for our sample to represent the population of interest. We use inductive reasoning to make inferences from our estimates, and we can make the best inferences when our estimates are unbiased and precise. Study design, sampling effort, and the estimators that we use are critical to provide reliable information from which to make decisions about wildlife populations.

References

Walther, B. A., and J. L. Moore. 2005. The concepts of bias, precision and accuracy, and their use in testing the performance of species richness estimators, with a literature review of estimator performance. Ecography 28: 815-829.

White, G. C., D. R. Anderson, K. P. Burnham, and D. L. Otis. 1982. Capture-recapture and removal methods for sampling closed populations. Los Alamos National Laboratory. LA-8787-NERP. 235 pp.

Williams, B. K., J. D. Nichols, and M. J. Conroy. 2002. Analysis and management of animal populations. Academic Press, San Diego.

Zar, J. H. 1999. Biostatistical analysis, 4th edition. Prentice Hall.

For more information on topics in this chapter

Conroy, M. J., and J. P. Carroll. 2009. Quantitative Conservation of Vertebrates. Wiley-Blackwell: Sussex, UK.

Cooch, E., and G. White. 2014. Chapter 1: First Steps. *In* Program MARK: a gentle introduction, 12th edition, Cooch, E. and G. White, eds. Online: http://www.phidot.org/software/mark/docs/book/pdf/chap1.pdf

Citing this primer

Powell, L. A., and G. A. Gale. 2015. Estimation of Parameters for Animal Populations: a primer for the rest of us. Caught Napping Publications: Lincoln, NE.

A nestling abbott's babbler (Malacocincla abbotti) is marked with an aluminum leg band, or ring, and a plastic, colored band for re-sighting in Khao Yai National Park, March 2003, Thailand. Photo by Andrew J. Pierce, used with permission.

Ch. 3

Maximum likelihood estimation[1]

"Every side of a coin has another side."
-- Myron Scholes

Questions to ponder:
- *What is a maximum likelihood estimator?*
- *No, seriously, what is a maximum likelihood estimator?!*

An introduction

Maximum likelihood estimation is a concept that is central to everything that we will discuss in this primer. So, there are some very good reasons for a student to learn the basics of maximum likelihood estimation. First and foremost, it is a phrase that will impress almost all of your colleagues, if you can use it in an appropriate sentence at the appropriate moment!

Unfortunately, maximum likelihood estimation is not well-understood by many who fear the quantitative explanations given by many textbooks. In fact, one of us (LP) was 90% of the way through his first parameter estimation course as a Master's student before he asked his officemate (a PhD student) if he could explain the phrase to him—they were using it in class all of the time, and he didn't know what it meant. Surprisingly, the PhD student couldn't explain it either. So, LP and his colleague sat down together, and looked for easy-to-understand explanations. Here, we'd like to take you through the same learning process that LP went through as a student. Maximum likelihood estimation is the most widely used method of parameter estimation—you need to learn the basics to appreciate the estimates you will eventually obtain.

We promise to take it slow.

If we break the phrase down into individual words, it may help with our understanding. Imagine that you have a sample of mark-recapture or survey data. And, imagine that you are trying to estimate the value of a parameter, such as density or the probability of survival. So, you could ask yourself, *"Given my data, what is the most likely estimate for the parameter?"*

---

[1] *With thanks for content to Therese Donovan.*

If you asked that question, you have just framed the concept of a maximum likelihood estimator. Using mathematics, **maximum likelihood estimation** is a method to **estimate the most likely value of a parameter, given a sample of data.**

Remember that we are operating "in the dark", so to speak. We do not know the true value of a parameter, but we do have our field observation data (capture records of marked animals, distance estimations of detected animals, number of sites occupied etc.). And, we can construct simple models of probabilities that we postulate to be true (these are our assumptions). So, the use of a maximum likelihood estimation method falls under the category of inductive reasoning: *if A and B are true and our sample is representative of our population, then it appears that the value of the parameter for this population is most likely X*.

Statisticians are good at helping us operate in the 'dark', without full knowledge of a population. If you value the input from smart statisticians, you'll be happy to know that maximum likelihood estimation methods are well-accepted, and can be traced back to Sir Ronald Fisher (*Fisher has been credited with providing the foundations for modern statistical science, as well as providing ground-breaking ideas for evolutionary biology*). Statistically, maximum likelihood estimators are unbiased, especially for large samples, and the variance provided by the estimator is minimized, especially for large sample sizes. Furthermore, maximum likelihood estimators are approximately normally distributed, and even non-statisticians are aware that we like to work with normal distributions when we conduct statistical tests.

> Statisticians tell us that the MLE method is easy to apply—we will let you decide for yourself after you finish this chapter. *If you find it easy, you may have found your calling as a quantitative ecologist!*

We can also make a list of reasons that are important to us, as biologists who work with quantitative methods, to become familiar with maximum likelihood estimation:

- Because maximum likelihood estimation is based on **likelihoods**, or **probabilities**, *the method is generally intuitive to us*—that is, we can make sense of it. In complex situations, we may agonize over the equations for probability statements (likelihoods), but the method allows us to 'work it out' in our brains.
- Likelihood methods are useful for parameter estimation, and we will also see (in the next chapter) that *likelihood theory is useful when we are conducting model comparison* with methods such as Akaike's Information Criterion.
- The use of maximum likelihood estimation *allows us to easily obtain estimates of variance* for our parameters.

Binomial coefficients

Most of the estimators that we will employ in this primer are based on a basic concept—we can establish probabilities associated with samples in a **binomial trial**. That is, in a trial for which there are only two possible outcomes (success/failure, heads/tails, alive/dead, emigrated/not emigrated, etc.), we can use basic mathematical facts to establish probabilities for the occurrence of an event (e.g., an animal living, or a head of a coin during a toss). So, we will begin our study of maximum likelihood estimation with the concept of the binomial coefficient to incorporate probabilities, or likelihoods. *Note: in latter chapters we will use* **multinomial** *coefficients when we have more than two possible outcomes such as when animals may be located in >2 areas during a survey or capture event.*

We use the expression of the binomial coefficient to calculate the number of ways, or the number of different combinations, a sample size of *n* can be taken from a population of size *N*. We express the binomial coefficient as:

$$\binom{N}{n}$$

And, we read it *"Big-N choose little-n"*.

For example, given a group of 3 individuals (N =3), how many combinations of two individuals (n=2) can be found? Generally, the formula for a binomial coefficient tells us that that we can calculate our answer as:

$$\binom{N}{n} = \frac{N!}{n!(N-n)!}$$

As a review, N! is read "N factorial". N! is the product of N*(N-1)*(N-2)*(N-3)*… until the last difference is 1.

Thus: 5! = 5*4*3*2*1=120.
And: 10! = 10*9*8*7*6*5*4*3*2*1 = 3,628,800.

So, to utilize the factorial statements in our example,

$$\binom{3}{2} = \frac{3!}{2!(3-2)!} = \frac{3 \cdot 2 \cdot 1}{2 \cdot 1(1)} = 3$$

Therefore, there are 3 combinations of two individuals that can be drawn from a group of 3 individuals. We can check the math with some visual logic, as our population is small and manageable. Let us give identifying letters to our 3 individuals: A, B, and C. We can see that it is possible to draw the following combinations of letters:

AB
AC
BC

So, there are three pairs or combinations, just as predicted. *Note that in this situation, the order of the letters does not matter. "AB" is the same combination as "BA".*

Let's do another example that may be a little more representative of something that we'll want to do for parameter estimation. Let's say that we are going to have 20 tosses of a coin—20 trials of a coin flip. How many combinations of coin tosses would result in exactly 5 heads (and thus, 15 tails)? We know we could start to write these out—the first three might be:

HHHHHTTTTTTTTTTTTTTT
THHHHHTTTTTTTTTTTTTT
TTHHHHHTTTTTTTTTTTTT
TTTHHHHHTTTTTTTTTTTT

You can see that this is going to take a while! And, the heads don't have to be all in a row, either...so, we better use math rather than just writing it out. Math can save us a lot of time!

We take our general formula and put in our values for N=20 and n=5. Give it a try. You should find that there are 15,504 combinations of heads and tails, from first toss to the 20th toss, that result in 5 heads. Aren't you glad we used math instead of writing them out?!

To make this 'biological', we could think of a radio-telemetry project with 20 bighorn sheep (*Ovis canadensis*). If we can imagine an experiment ending with 5 of the sheep alive after a year, how many combinations of sheep (sheep #1, #6, #7, #10, and #18, for example) could we have that allow 5 sheep alive?

Of course, the answer is the same as the example with the heads and tails: 15,504.

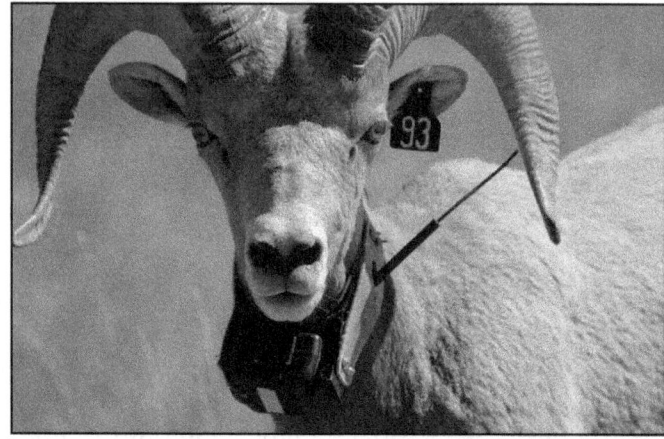

Figure 3.1: *Radio- and ear-tagged male bighorn sheep in Nebraska (photo by Kelly Powell, used with permission).*

This approach—enumerating possible combinations from our samples—will be important when we use maximum likelihood estimation to estimate survival rates. But, for now, we can be happy that we are able to use the binomial coefficient to think about samples.

MLE: a fish tagging example

Let's illustrate the concept of maximum likelihood estimation with an example. Imagine that biologists are conducting a tag loss study of fish—before embarking on a larger study, the scientists want to know if their fish tags will stay in place or, instead, will simply fall off, which would be bad for the larger study!

So, the biologists put 10 fish in a tank, and each fish is marked with the special tag (Figure 3.2). After one day, the animals are checked. Seven of the 10 fish have retained their tags, and three of the fish have lost their tags. The biologists need to estimate the probability of tag retention over the 1-day period

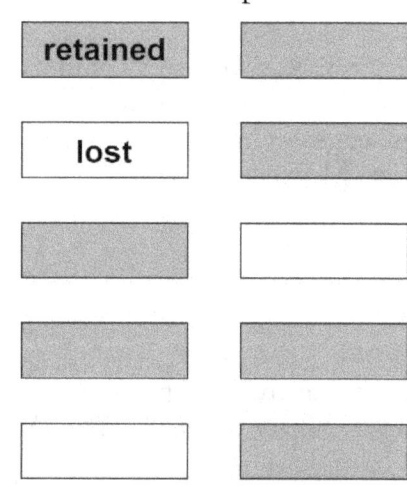

Figure 3.2: *Depiction of 10 marked fish placed in a tank. Shaded boxes represent fish that retained tags, and open boxes represent fish with lost tags.*

of time. And, they are pretty sure they should use a maximum likelihood estimation process.

However, the 12-year-old daughter of one of the scientists, Dr. Bigstuff, enters the laboratory and tells them they are thinking too hard—the answer is that the tag retention probability is 70%. All of the other scientists agree—she appears to be correct. It's obvious—you use the fraction 7/10. Seven of 10 fish remained tagged. That's 70%. Case closed, right?

Instead, Dr. Bigstuff wants to prove that his daughter is not correct. He still thinks a maximum likelihood approach would be best for this important project. He sends his daughter home to work on her mathematics homework, while he sits down to work on his own math problem.

In the above section, we learned about **binomial sampling**. We'll put that information to use here as we help Dr. Bigstuff.

> *A **binomial trial**, also known as a **Bernoulli trial**, has exactly **two outcomes**. The trial will end in either a success or a failure. When we apply this to our fish tagging, we can label the tag retention as a success and a tag loss as a failure.*

We should point out that we could also label the tag loss as the success if that made more sense (it doesn't seem to, does it?!)—but we must be clear about the definition as we move forward, and we must remember which outcome we labeled as a success.

In an experiment based on binomial trials, we know that we will have n trials and we will have y successes. And, we know that y is an integer and the value of y can be stated as: $0 \leq y \leq n$.

Now that we have labeled our success (tag retention), we can identify a parameter of interest. In binomial trials, we usually use p to represent the probability of success. And, conversely, q is the probability of failure. If there are only two outcomes of our trial, then we know that the following is true: $q = 1-p$

And, because p is a probability, p will be continuous with the value of: $0 \leq p \leq 1$

Last, because we have data, we can estimate p with a maximum likelihood estimator. Maybe Dr. Bigstuff is correct…?

Writing the likelihood statement

A binomial trial is a very simple experiment. And, as a result the "likelihood statement", or "probability statement", is also relatively simple. The general Bernoulli likelihood formula is written as follows:

$$L(y \mid N, p) = \binom{N}{y} p^y (1-p)^{(N-y)}$$

Formulas are nice, but it's good to be able to translate them into words—so, let's do that. On the left side of the equation, above, is the statement, $L(y|N,p)$. This can be read, "the likelihood (or probability) of observing y (the number of successes), given N (the number of trials) and the presence of a parameter, p, equals…" Another way to say this is, "We'll be estimating the value for a parameter given some data and underlying assumptions about a simple model." That should sound like our definition for maximum likelihood estimation…

On the right side of the equation, we have a more complex statement. You should recognize the binomial coefficient:

$$\binom{N}{y}$$

We use the binomial coefficient to calculate how many ways (combinations) you could get what you are proposing (in our case: *how many ways could 7 tags be retained on 10 fish?*).

The product

$$p^y(1-p)^{(N-y)}$$

is the probability statement that will estimate the probability of getting our exact result (seven successes and three failures in our fish tagging experiment) in any order (could be Fish #1-7 retain tags, or it could be Fish #2-8 retain tags, etc.). If p is the probability of one success, we need to raise p to the yth power to get the probability of y successes. And, if $q=1-p$ (representing the probability of a failure), we need to raise $1-p$ to the $N-y$ power to get the probability of $N-y$ failures.

> In our fish tagging example:
> $y = 7$ successes
> $N-y = 3$ failures

And, for our specific experimental results with 10 trials and y observed successes, Dr. Bigstuff could arrange the equation (here we use $L(\theta|y=7)$ to represent $L(y|N,p)$ and to designate the specific results of our experiment: $y=7$ successes) as:

$$L(\theta \mid y = 7) = \binom{10}{7} p^7 (1-p)^{10-7}$$

We now have our **likelihood statement** and this is the first step along the path to use a maximum likelihood estimator. Remember that we said the MLE process would be logical, because we are using probabilities? And, our fish tagging example shows the relatively simple pieces that go together to construct a likelihood statement.

Maximum likelihood cookbook

Luckily for most of us, we will never have to do MLE's by hand. That is why we paid a lot of money for our computers, right?!

But, for the sake of understanding what is going on when your computer obtains a MLE, there are two basic steps in the MLE process:
- The **first step** is to **state the structure of our model and write out the likelihood function** for our experiment. We've done that for the fish tagging experiment. We had 10 trials, and it was a binomial situation with a success or a failure in each trial. And, we

wanted to estimate a parameter—the probability of success in the trial (p). Done. On to step two.

- The **second step** involves *calculus. Please don't close the book.* It's not that painful. To get the maximum likelihood estimate, we must **maximize the likelihood function**. Whoops! Did you notice that? Probably not—you were worried about the calculus. But, there was a word that we were looking for: ***maximum***. By maximizing the likelihood function, we will be able to determine what value of the parameter is most likely, given our data. It is all coming together—really!

> **Calculus** is a mathematical approach to study either slopes and curves (differential calculus) or areas under curves (integral calculus). In this chapter, we will use basic concepts of differential calculus. In Chapter 19, we will use concepts of integral calculus.

Maximizing the likelihood: with graphics

Let's go back to our example of 7 fish that retained their tags in an experiment of 10 tagged fish. Again, we start with this likelihood, for our 10 trials and 7 successes:

$$L(\theta \mid y = 7) = \binom{10}{7} p^7 (1-p)^{10-7}$$

If we simplify the right side of the likelihood function, we get the following (worked out in steps here):

$$L(\theta \mid y = 7) = \left[\frac{10!}{7!(10-7)!}\right] p^7 (1-p)^3$$

We can simplify this by writing out the factorial statements. If $10! = 10*9*8*7*6*5*4*3*2*1$, and if $7! = 7*6*5*4*3*2*1$, then $10!/7!$ in the above equation can be simplified to $10*9*8$. Then:

$$L(\theta \mid y = 7) = \left[\frac{10*9*8}{3*2*1}\right] p^7 (1-p)^3$$

And finally:

$$L(\theta \mid y = 7) = 120 p^7 (1-p)^3$$

Our "cookbook" directions above tell us that we now need to take our likelihood function and "maximize" it. There is a reason that calculus is needed at this step—it can help us find the maximum of the function that we've written.

Before we go further, we should point out that a function is simply the relationship between two things—in our case the function describes how the value of the likelihood (L) changes as p

changes in value. We can simplify the left side of the equation to help us visualize this equation better:

$$L = 120 p^7 (1-p)^3$$

And, we can graph that function by replacing p with various values between 0 and 1 (the range of possible p's (Figure 3.3).

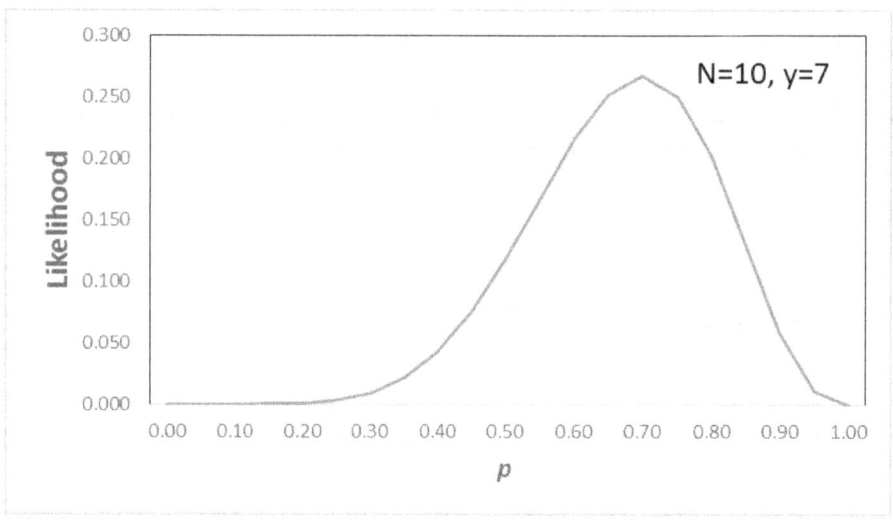

Figure 3.3: *The value of the likelihood function, $L=120p^7(1-p)^3$, for a range of values of p. The function reaches its maxima (peak) at $p=0.7$.*

You can try this in a spreadsheet yourself. In a blank spreadsheet, make one column labeled "p" and a second column labeled "Likelihood".

Under the "p" heading, put a column of numbers like follows:

0
0.05
0.10
...
0.95
1.00

Then, in the likelihood column, create an equation to calculate the likelihood. If you are using Microsoft Excel, the equation (in Excel format) would look something like this for the top cell (with $p = 0$ in the cell to its left).

= 120*(A2^7)*(1-A2)^3

If you copy that formula to the rest of the cells below in the B column, you should see values that match the ones to the right. And, you can use your Excel graphing skills to see if you can replicate Figure 3.3!

p	Likelihood
0	0
0.05	8.03789E-08
0.1	8.748E-06
0.15	0.000125915
0.2	0.000786432
0.25	0.003089905
0.3	0.009001692
0.35	0.021203015
0.4	0.042467328
0.45	0.074603106
0.5	0.1171875
0.55	0.166478293
0.6	0.214990848
0.65	0.252219625
0.7	0.266827932
0.75	0.250282288
0.8	0.201326592
0.85	0.129833721
0.9	0.057395628
0.95	0.010475059
1	0

Now that we've graphed it, can you see where the maximum of the function is? What value of p makes L (the likelihood) largest? *Interestingly, it looks like it is somewhere near $p=0.7$.* Perhaps Dr. Bigstuff should start listening to his daughter?!

Well, Dr. Bigstuff might claim that we can't tell exactly where L is the highest—perhaps it is at $p=0.68$ or $p=0.72$? We only used values that differed by 0.05 in our graphical approach. What is the <u>exact</u> maximum?

Maximizing the likelihood: with calculus

To find the exact maximum of our likelihood function, we must turn to calculus. We know you've repressed everything you learned in that class and you may not remember much…but, do you remember that if you take the **derivative** of a function, it tells you the slope at a given point along the function? Maybe you don't remember that, but take our word for it. It's true.

Using that concept, we can find where the steepest part of the curve is (that part of the curve would have the largest value for the slope), and we can find where the slope is not very steep. For example, if we draw a tangential line to the function where $p=0.6$, what is the slope?

That would be line A in Figure 3.4, and we can see the slope is positive and very steep.

Next, we can ask ourselves—*what would be the value of the slope at the <u>maximum</u> point (peak) of the curve?*

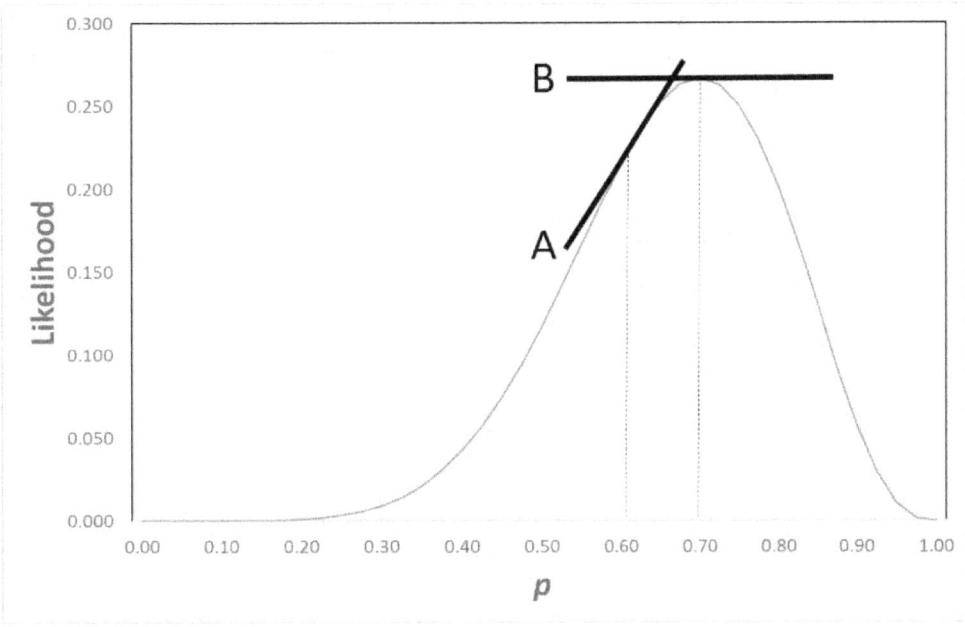

Figure 3.4: *The likelihood function shown in Figure 3.2 with depictions (A) of the slope of the function at a given value of p (here, 0.6) and (B) the slope of the function at the maxima (peak) of the function ($p=0.7$). At the peak, the slope=0 (no rise; flat line).*

Aha! The correct answer is—***the slope would be zero***! We can see this by looking at line B in Figure 3.4. The slope at the tip of the curve is completely flat.

Importantly, we can also notice that the maximum is the ONLY point on the curve with a slope of zero. That's important, as well.

Thus, we can use logic and put together a powerful idea—if the derivative of the likelihood function gives us its slope, and if the slope at the maximum point of the likelihood function is zero, then we should be able to set the derivative of the likelihood function equal to zero and solve for the value of p at which the maximum occurs. This is the crux of maximum likelihood estimation!

We can now celebrate—we've used calculus for something useful! Let's apply this to our fish tagging example. Note that we identify the derivative of a function (e.g., Z) as Z'. So, we can write our likelihood again, as before:

$$L(\theta \mid y = 7) = 120 p^7 (1-p)^3$$

Then, we indicate that we need the derivative of the function, and we set it equal to 0.

$$\left(120 p^7 (1-p)^3\right)' = 0$$

Now, we need to calculate the derivative. If you have retained a lot from calculus class, you may be able to figure this out. If not, we suggest you can quickly find the answer by googling "on-line derivative calculator" and using the on-line tool to get your answer, which is as follows (trust us, or do it on-line to see if we're correct!):

$$840(1-p)^3 \cdot p^6 - 360(1-p)^2 \cdot p^7 = 0$$

We can simplify that result as follows:

$$\frac{840}{360}(1-p) = p$$

And, then:

$$2.333 - 2.333 p = p$$

And, further:

$$2.333 = 3.333 p$$

Now, if we solve for p, we find that **the value of p that maximizes the likelihood function is 0.7.** Thus:

$$\hat{p} = 0.7$$

Thus, Dr. Bigstuff's daughter was correct. Her simple fraction of $p = 7/10$ or $p = y/N$ was truly a maximum likelihood estimator! See, you have been using maximum likelihood estimation since elementary school, and you didn't know it.

Coin toss example

Let's do another example.

> **A little hat?**
>
> Throughout this primer, you will see parameters, such as p, indicated as p or as $\hat{p}$. The "hat" symbol indicates that this is an **estimate** of p rather than a theoretical value or a known value of p.

In the section above, we worked with an example of 20 tosses of a coin. And, we asked about the probability of getting 5 heads and 15 tails in such a trial. This is very similar to our experiment that we've just completed with the fish and the tags! So, let's see if we can apply our knowledge to this new question.

Our coin flipping scenario is similar to the fish tagging experiment—we have a certain number of trials (20 coin flips) and we want to have 5 successes (flipping a "head", Figure 3.5). So, we can use the same likelihood formula for a Bernoulli trial:

$$L(5 \mid 20, p) = \binom{20}{5} p^5 (1-p)^{(20-5)}$$

We're using this example as we know that the probability of flipping a "head" with a "fair coin" (not weighted in any way) is 50% or p=0.50. If we replace the p's in the formula with 0.5, we find that the last portion of the equation

$$p^5(1-p)^{(20-5)} = 0.00000095$$

This is the probability of getting *one* of *any* combination of 5 heads and 15 tails in 20 flips, without worrying about the order.

That is a really small number (0.00000095), but we also know that we need to account for all of the combinations of ways that we could get 5 heads and 15 tails. So, we look to the first portion of the equation. As earlier in this chapter, we find:

$$\binom{20}{5} = 15504$$

Thus, there are 15504 ways to have 5 heads and 15 tails in 20 flips of a coin. And, each has a small chance of occurring *(0.00000095, to be exact!)*. So, we need to multiply the two portions of this equation together—the total probability of getting some combination of 5 heads and 15 tails is equal to the number of combinations possible, multiplied by the probability of getting any one combination of 5 heads and 15 tails.

Figure 3.5: *A specific result of trials of 20 tosses of a coin referenced in the text: 5 heads and 15 tails.*

We rely on the formula to tell us that **if $p = 0.5$, the likelihood of getting exactly 5 heads in 20 coin flips is 0.0148.** Or, to say it another way, if we performed many, many trials of 20 coin flips, we'd expect to get 5 heads in 1.48% of our trials. And, this makes intuitive sense to us. Most of the time, we would probably get about 10 heads and 10 tails, because our coin is not defective (we assume, if $p=0.5$). So, getting only 5 heads should be a rare occurrence.

Proving why we need both portions of the likelihood:

Pretend for the moment that our coin flip example is simpler: we have 3 flips of a coin and we want to know the probability, or likelihood, of getting 2 heads and 1 tail.

Our likelihood expression would be: $L = \binom{3}{2} p^2 (1-p)^{(3-2)}$

The first part of the equation tells us that there are $3*2*1/2*1=3$ combinations that can give us 2 heads: HHT HTH and THH.

The second part of the equation tells us that the probability of getting one of ANY combination of coin flips with 2 heads and 1 tail is $0.5^2(0.5)^1 = 0.125$.

But, we can get 2 heads and 1 tail by 3 different combinations, so we multiply the two portions together, and we find: $3*0.125=0.375$. So, we have a 37.5% chance of getting 2 heads and 1 tail in 3 flips of a coin.

Because this is a small trial, we can test our answer by writing out all possible results of 3 coin flips:

    HHH—*only 1 way to get 3 heads*
    HHT—*three ways to get 2 heads*
    HTH
    THH
    HTT—*three ways to get 1 head*
    TTH
    THT
    TTT—*only 1 way to get 0 heads*

There are 8 possible results when flipping a coin 3 times. And, only three of them give us 2 heads! By simple math, $3/8 = 0.375$, or 37.5% of the results. It works.

Meanwhile, back in the real world

We've just explored the use of a likelihood statement for which we know the value of *p*—because we know that a 'fair coin' should give us a head 50% of the time, on average.

But, we don't know the value of *p* when doing a study in nature. Let's change the "story" without changing the numbers—let's pretend that we release 20 marked animals and only 5 survive (Figure 3.6).

What is p, the probability of success (or the probability of survival)?

Figure 3.6: *A sample of 20 tagged animals in which only 5 survived. Dead animals are marked with an "X".*

Of course, we can use MLE to tell us the value of *p* that would maximize the chances of our observation occurring. That is, *given our observation of 5 animals surviving from an initial cohort of 20, what value of p is most likely?*

We still have the same formula for the likelihood:

$$L(5 \mid 20, p) = \binom{20}{5} p^5 (1-p)^{(20-5)}$$

It may help us to graphically view our function, over a range of *p*'s and see if we can identify the value of *p* that maximizes the function. If we do this in a spreadsheet such as Microsoft Excel, we can see that there certainly is a peak—a maxima—to the function that we plot. What is the value of *p* at the maximum point? Yes, it appears to be at or near *p*=0.25 (Figure 3.7).

In our previous fish tagging example, we looked at a graphic to see where the likelihood function appeared to have a maxima. Then, we took the likelihood equation with our specific values (7 tags retained out of 10 fish), and we worked out the value for *p*. We could do the same here for our 20 animals released and 5 survivors.

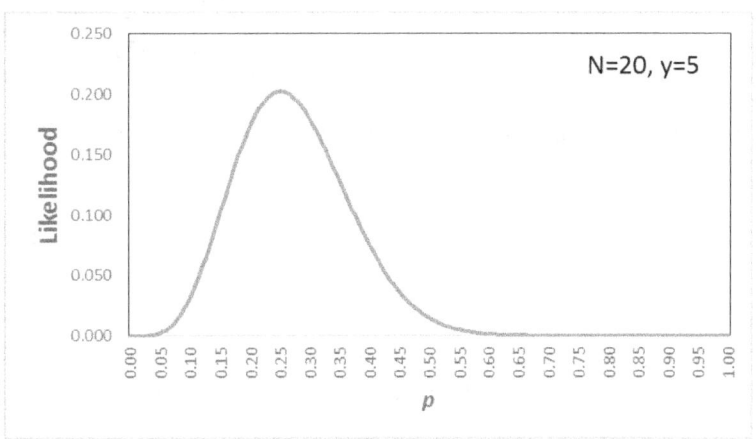

Figure 3.7: *The value of the likelihood function for the study and sample results depicted in Figure 3.6.*

But, we might pause and ask ourselves—*are we going to have to do this set of calculations for every single observation in every experiment we ever use?* We hope not! It would be nice to have a general formula—an **estimator**—that would give us our estimate.

Luckily, statisticians have done some preparatory work for us. They have provided an elegant way to obtain a maximum likelihood estimator for a simple Bernoulli experiment with N trials and y successes.

If we take our general Bernoulli likelihood statement,

$$L(y \mid N, p) = \binom{N}{y} p^y (1-p)^{(N-y)}$$

and then take natural logarithms of the entire likelihood, we obtain the "log likelihood".

Mathematicians suggest doing the log transformation to the equation because it simplifies the analytical process—we remove the exponents when we take the log of each side of the equation to get:

$$\ln L(p \mid data) = y \ln p + (N-y) \ln(1-p)$$

Then, we take the derivative of the equation, with respect to p. Again, this provides us with an equation that can give us the slope of the function at any value of p. We are most interested in the value of p where the slope is zero (the maxima!), so we set the derivative equal to 0:

$$\frac{\delta[\ln L(p \mid data)]}{\delta p} = \frac{y}{p} - \frac{(N-y)}{(1-p)}$$

$$\frac{y}{p} - \frac{(N-y)}{(1-p)} = 0$$

Then, we simplify the equation to give us our maximum likelihood estimator for p:

$$\hat{p} = \frac{y}{N}$$

Now, if we go back to our 20 animals with 5 survivors, we find that $\hat{p} = 5/20$, or 0.25. That matches our best guess from looking at Figure 3.6, correct?

These are fairly simple examples of maximum likelihood estimators, using a simple Bernoulli trial. Although it is simple, this is actually the estimator for known-fate survival that can be obtained from radio-marked animals! Scientists release a certain number of animals and have a certain number of survivors every day somewhere out in the field. So, even though we started with a simple example, it is useful, and you will see this concept again in later chapters.

Variance estimation, MLE-style

Before we leave the topic of maximum likelihood estimation (although we might ask in a metaphysical sense, do we <u>ever</u> really leave the topic of maximum likelihood estimation?!), we need to talk about variance estimation. This is important—there is no journal on earth (well perhaps a *slight* exaggeration) that will allow you to publish a parameter estimate without an estimate for the variance, the standard error, and/or a 95% confidence interval!

Luckily, if we look back at the benefits of using maximum likelihood estimation methods, we see that one advantage is that it is straightforward to estimate the variance for a parameter using MLE.

Let's start by investigating this graphically (Figure 3.8):

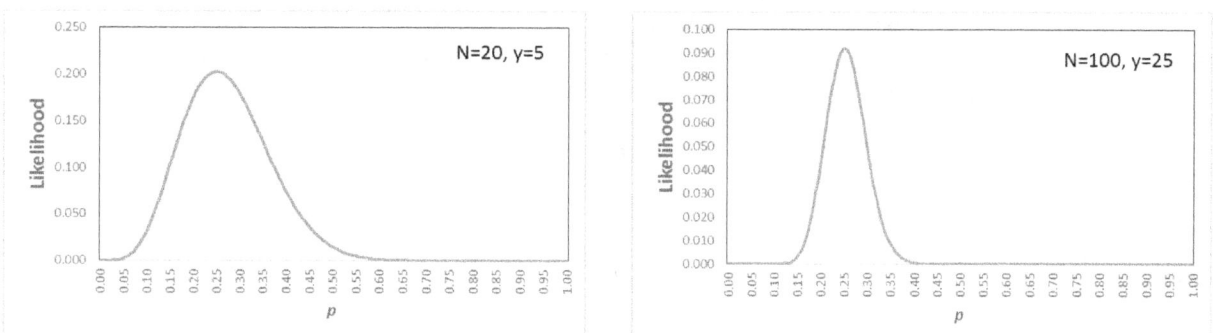

Figure 3.8: *Comparison of two similar likelihood functions, both with maxima at p=0.25. The function at left represents a study with a sample size of 20, while the function at right represents a study of 100 tagged animals.*

You'll notice that both figures appear to have maxima at $p=0.25$ (Figure 3.8). For reference, the figure at left is for the example we just covered—a simple Bernoulli trial of 20 animals tagged with 5 survivors and survival was estimated as $\hat{p}=0.25$.

The difference in the curves in the two graphs (Figure 3.8) is that the one on the right is from a larger experiment. This scientist had more money for radio-tags! They released 100 animals and 25 survived. If we put those numbers into our estimator for survival, we find that again, survival was estimated as 0.25.

So, the same model structure (a Bernoulli trial) and the same estimate of the parameter. The only thing that differs is the sample size, or the number of trials (N).

Conceptually, **sampling variance** is related to the curvature of the likelihood function at is maximum. At the peak, or the maxima, is the curvature narrow or wide? It is wider on the figure to the left and narrower on the figure to the right.

We know from elementary statistics that sample size (N, the number of trials) affects the sample variance of the estimate, $\hat{p}$. We see this graphically in Figure 3.8—clearly, the sample size changes the relative likelihood of the function at various values of p. A narrower range of values for p have high likelihood values (and thus, a much narrower range of values of p are likely, given the observed data and the model structure).

Mathematically, we have to go back to calculus for the variance estimation. We've already taken the derivative of the likelihood and set it to zero to find the value of p that maximizes the likelihood function. We could call that derivative the **first derivative** because it is ...*well*... the first derivative that we calculated for that likelihood.

We could also take the derivative of the first derivative (let's see...perhaps we should call it the **second derivative**?!). In calculus, the first derivative gives us information about the slope of the line (steep or shallow or flat), and the second derivative gives us information about the curvature of the line. Based on what we described above about the curvature at the maxima for the likelihood, it sounds like the second derivative of our likelihood function might be useful!

And, it is.

Specifically, if we trust the estimable Sir Ronald Fisher to guide us, we are told that the negative inverse of the second derivative provides the maximum likelihood estimate for the variance, and for our simple Bernoulli trial, the variance estimate is:

$$\mathrm{var}(\hat{p}) = \frac{p(1-p)}{N}$$

You try it

You can have some fun with this simple maximum likelihood estimator by placing 100 beans of 2 colors (any combination) in a cup. Mix a known number of dark and light beans (any seed pods or other small colored objects will work). So, in this example you know the truth about p, the proportion of light beans (a success).

Now, take a series of samples—first take 10 beans at random from the cup. $N = 100$ and $y = 10$ (the number of light-colored beans you get in your sample). What is your estimate for p? Is it close to what you know to be the truth?

Now, use the following formula to estimate the confidence interval for your p-hat:

Lower bound of 95% confidence interval: $\hat{p} - 1.96 \cdot \sqrt{\frac{p(1-p)}{N}}$

Upper bound of 95% confidence interval: $\hat{p} + 1.96 \cdot \sqrt{\frac{p(1-p)}{N}}$

Does the 95% CI contain the true value of p? Perhaps you need a larger sample to reduce bias and increase precision? So, put the 10 beans back in the cup with the other 90 beans and mix them up. Now, remove 25 beans and estimate $\hat{p}$ and calculate the 95% CI again. Last, try removing 50 beans as your sample. N will always equal 100, but y will change: 10 to 25 to 50.

What happens to your confidence interval? It should get smaller as your sample size increases.

Conclusion

Maximum likelihood estimates will be used 'behind the scene' in the methods described throughout this primer. A basic knowledge of the process used to establish a maximum likelihood estimator is valuable for a biologist to have. If nothing else, remember that a maximum likelihood estimator is a method to find the most likely value for a parameter, given the model structure and our observed data.

To find the maximum likelihood estimate, two steps are necessary: (1) you must write the statement of likelihood for your experiment (our examples were Bernoulli trials), and (2), you must use the likelihood function and a small (*really!*) amount of calculus to find the maxima of the likelihood function.

For more information on topics in this chapter

Conroy, M. J., and J. P. Carroll. 2009. Quantitative Conservation of Vertebrates. Wiley-Blackwell: Sussex, UK.

Cooch, E., and G. White. 2014. Chapter 1: First Steps. *In* Program MARK: a gentle introduction, 12th edition, Cooch, E. and G. White, eds. Online: http://www.phidot.org/software/mark/docs/book/pdf/chap1.pdf

Donovan, T. M. and J. Hines. 2007. Exercises in occupancy modeling and estimation. Online: http://www.uvm.edu/envnr/vtcfwru/spreadsheets/occupancy.htm

Williams, B. K., J. D. Nichols, and M. J. Conroy. 2002. Analysis and management of animal populations. Academic Press, San Diego.

Citing this primer

Powell, L. A., and G. A. Gale. 2015. Estimation of Parameters for Animal Populations: a primer for the rest of us. Caught Napping Publications: Lincoln, NE.

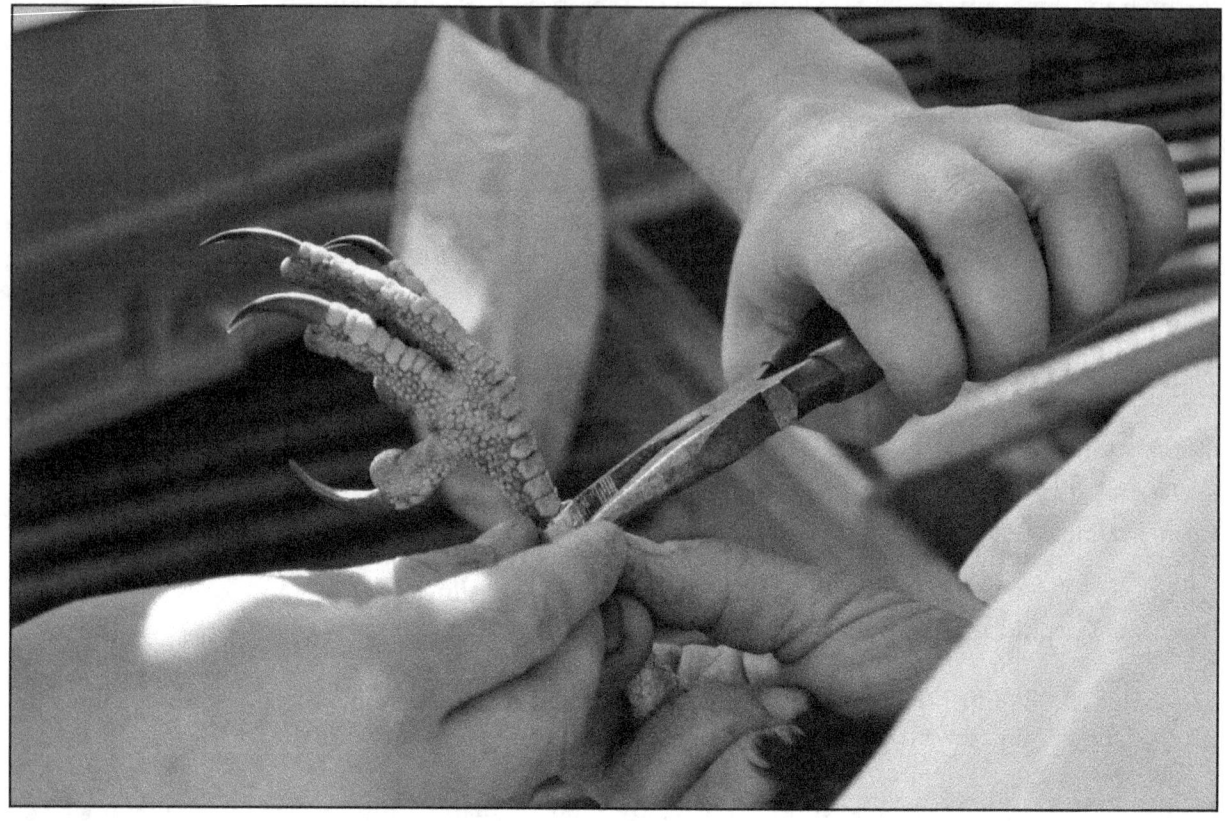
A northern crested caracara (Caracara cheriway) is tagged with an aluminum leg band in Florida, USA. Photo by Jennifer Smith, used with permission.

Ch. 4

AIC and model selection[1]

"There are only two mistakes one can make along the road to truth; not going all the way, and not starting."
-- Buddha

"If you are out to describe the truth, leave elegance to the tailor."
-- Albert Einstein

Questions to ponder:
- *What does a "model" describe, in the context of parameter estimation?*
- *Do we know which model is the best model to use?*
- *What is AIC, and how is AIC calculated?*
- *Should I use AIC_c or AIC?*
- *How do I interpret AIC values?*
- *What is model averaging?*
- *Do I always need to model-average my parameter estimates?*
- *It seems like everyone uses a different method for model averaging. What should I do?*

Start here, but keep going

A discussion of model comparison is critical to the background section before we begin to explore the various types of parameter estimation processes. Here, we will provide a cursory introduction. To continue your journey, we point you towards the gold standard for ecological investigations and model inference: Burnham and Anderson (2002). And, we heartily recommend another small primer, Anderson (2007) that picks up where we leave off with an easy-to-understand approach.

This chapter has two parts: first, an introduction to the theory and concept of AIC and model comparison. Then, we finish with a discussion of the process (which can often be complex!) of selecting the "best model" and how to present your parameter estimates to the world.

---

[1] *With thanks for content to Evan Cooch and Gary White.*

Part One: Multi-model inference

A person does not need to use multi-model inference to estimate the value of a parameter. In fact, in all of the examples used in this primer to this point, we have looked at one model and estimated the value of a parameter based on that model.

But, ecology is complex. Most biologists estimate parameters (e.g., survival, density, movement probabilities) in the context of a question. Does survival vary between genders or ages? Does density vary among my study plots? Which management scenario results in the highest probability of nest survival for mallard ducks (*Anas platyrhynchos*)?

To answer these questions, ecologists have long-used an approach of alternative hypotheses (Chamberlin 1965). We use **"model"** in many ways (see Chapter 1), and one way is to describe a concept (or hypothesis, in this case) for the truth with respect to values of parameters for a population. One "model" might suggest that survival is the same among all animals, while another "model" would propose that survival varies by gender. Still another "model" might suggest that survival, instead, varies by age of the individual (e.g., juvenile and adult categories).

The statistical approach to alternate hypotheses regarding demographic parameters has changed over the years—the earliest, simplest form was to compare a null model with one alternative model in which a parameter varied according to categories of animals. For, example, do annual survival rates of northern bobwhite (quail [*Colinus virginianus*], Figure 4.1) differ by gender (male and female categories)? Perhaps age (juvenile and adult categories)? We could lump our sample into data obtained from males and data obtained from females, and we could estimate a survival rate for each group. Hines and Sauer (1989) provided a comparison using program CONTRAST to assess whether the resulting estimates of survival (and its associated estimate of variance) provided evidence of a difference. If there was no evidence for variation of survival according to gender, the null model of "no difference" would be supported. At that time in history, the ability to make simple comparisons was a huge step forward in our analyses. Today, it seems very simple (*which, we remind the reader, does not mean it is useless!*).

One problem with the approach used by CONTRAST was that it was limited to categorical variables (e.g., gender or age) to describe differences in a parameter. What if a model suggested a continuous variable might be important—for example, might the survival of quail change with respect to body mass (small birds have lower survival than large birds?)? Program CONTRAST couldn't handle that comparison.

As a result, since the mid-1990's, ecologists have developed easy-to-use methods to allow exploration of categorical (discrete) or continuous effects on the value of parameters. And, during this time, ecologists have increased their use of an analysis framework known as multi-model inference, which uses model comparison as its basic structure. The general idea is that a team of scientists can put forward two or more models that they

Figure 4.1: *Northern bobwhite (Colinus virginianus). Photo by BS Thurner Hof, available in the public domain.*

believe to represent alternative, biological hypotheses about the way the world "works" with respect to the value of the parameter(s) they are estimating. Each model, individually, is constructed. The values of the parameters are obtained. Then, the models are compared with each other to determine which model best represents variation in the value of parameters that is supported by the data.

So, for example, a team of quail biologists could submit the following models about annual survival of quail in northern Florida:

Model	Description
Null model	The entire population has the same probability of annual survival
Gender model	Males and females vary in their probability of annual survival
Age model	Juveniles (<1 year) and adults vary in their probability of annual survival
Mass model	Mass (g) of an individual affects the probability of annual survival

So, the biologists have now created four models and are ready to compare them. Before we let them continue with their comparison, we need to pause and think about the philosophical framework.

In Chapter 2, we discussed inference methods as ways in which scientists seek "truth" about their biological systems. So, philosophically, there is some "truth" about our population, and we are trying to get as close as we can to the "truth." But, we have also acknowledged that ecologists do not know (and may never know) the complete truth about their biological systems. Thus, you might ask: *"How will we ever know which of our models is the best model, if we don't know the truth?"*

And, it is true—if we knew "truth", it would be easy to compare the results from our parameter estimation to "truth". It seems fairly easy to understand that a side-by-side comparison of all of our models to "truth" would show which model is "best" at describing the system. In fact, statisticians have named the theoretical distance between any model and the full reality, or truth, the **Kullback-Leibler distance** (Figure 4.2). In information theory, the size of that distance provides information, termed

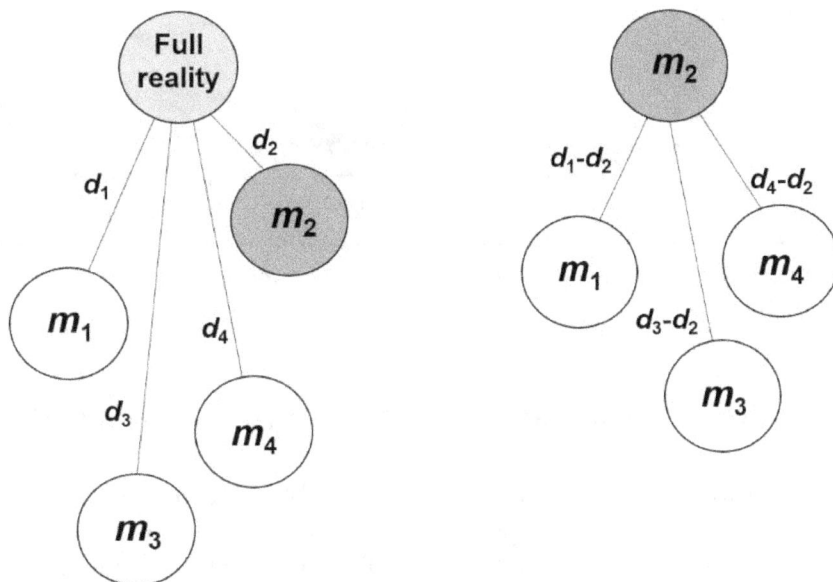

Figure 4.2: *The theoretical distances (d_i, Kullback-Leibler information) between full reality and competing models (m_i) are shown at left. The estimated distances from the best model (m_2) to other models provide the conceptual framework for AIC values and ΔAIC for model comparisons (after Cooch and White 2014).*

Kullback-Leibler information. We could use that information to compare the relative merits of our models.

But, we don't know the truth.

Luckily, statisticians have also theorized that if such "real" distances between models and truth exist for our set of models, then we can estimate the distance from all models to the best model. To understand this, consider two cities (City A and City B) that exist near a sacred mountain with a sacred population of mountain goats (*Oreamnos americanus*) at its top in a faraway land. Both of the cities can see the mountain (and on a day with good visibility, the goats). And, citizens of both cities know that City A is closer to the mountain. There is direct evidence of which city is closest to the holy site (Figure 4.3).

Our story takes a turn for the worse when a massive earthquake occurs. The mountain disappears, the sacred mountain goats all perish, and only a few people in City A and City B survive. Over generations, the cities rebuild. Eventually, a shopkeeper in City A finds a book with stories of a sacred mountain that once stood near the cities. The book contains the phrase, "Blessed will be the city that is closest to the sacred mountain..." But, the rest of the book was destroyed in the earthquake, and no one knows which city is closest. Without a reference point, it is impossible to know which city was the blessed city. Or...is it?

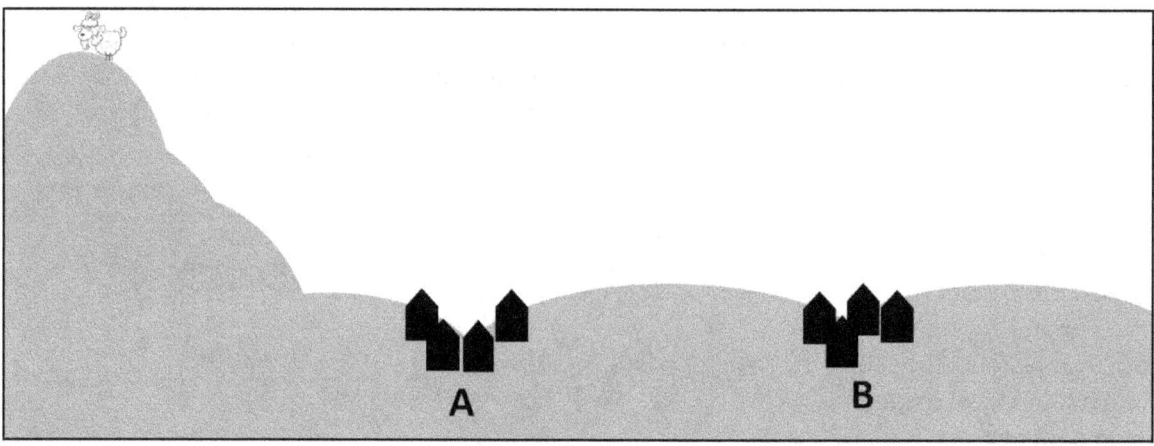

Figure 4.3: *Relative distances from City A and City B to the sacred mountain, the similarly sacred sanctum for a population of mountain goats.*

Akaike's Information Criterion to the rescue

Many readers will confirm that it is impossible to read many ecological papers that deal with parameter estimation without seeing a reference to **Akaike's Information Criterion**, or **AIC**. And, it is not difficult to determine—from the ecological papers that use AIC values—that AIC is a method of model comparison. That is, AIC allows us to compare our models in attempt to select the best model among our model set.

What exactly is Akaike's Information Criterion? AIC was developed by the Japanese statistician, Akaike, as an unbiased estimator of the "relative, expected" Kullback-Leibler distance. Thus, AIC is a method that identifies which model is closer to the truth that appears to be expressed by the data from your sample.

So, without knowing the location of the mountain, the AIC model comparison concept allows us to use evidence to learn about City A and City B (Figure 4.3). Perhaps, after searching through all known diaries of residents of the two cities, we learn that more diaries from City A mention the mountain. And, more holy relics are found in City A. As evidence builds, we become more and more certain that City A was closest to the mountain. So, we use evidence to determine that A was most likely the closest city to the original site of the sacred mountain without knowing where the mountain once stood. That's spooky, but it is pretty useful! And, that is essentially what the AIC method does—it assesses evidence that is gathered in our sample.

Conceptually, Akaike's method addresses a trade-off between model fit (reduced bias) and the variance of the estimate, and this trade-off is important (Figure 4.4). On one hand, we want a model that 'fits' (adequately explains) the variation we believe is in our sample data. Such an approach would favor a more complex model—one with more parameters—to attempt to explain variation. For example, we might believe that survival of quail is affected by gender and age of the bird. So, we'd need a complicated model to estimate survival for juvenile males, juvenile females, adult males, and adult females (4 parameters estimated).

On the other hand, as we estimate more parameters with the same pool of data, we create more variance components and more uncertainty about the value of each parameter (we address why this is true with an example below). We like precision, so we would want to have a simple model with fewer parameters and lower variance. A model that uses all the data to estimate one survival rate (no age or gender effects) would be the simplest and would give a survival rate with a lower standard error. AIC addresses that trade-off to determine if we have enough evidence to support a more complex model.

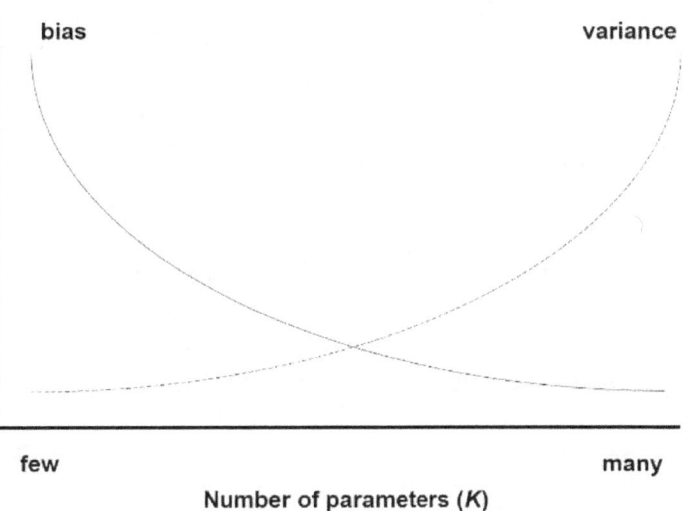

Figure 4.4: *Conceptual trade-off between model fit (high bias = poor fit) and uncertainty for a parameter estimate, as expressed by variance of parameters in the model. Models with few parameters may have more bias than highly parameterized models, but models with many parameters may have high estimates of variance for the parameter values (after Cooch and White 2014). Y-axis is scaled "low" (at origin) to "high" for both bias and variance.*

AIC: "fit" versus variance components

To understand the idea of complexity in a model, relative to evidence to support such a model, let's step away from estimation of survival rates and think about a population's trend over time. Let's suppose a biologist gathers information on population size from surveys every year, resulting in data points displayed in Figure 4.5.

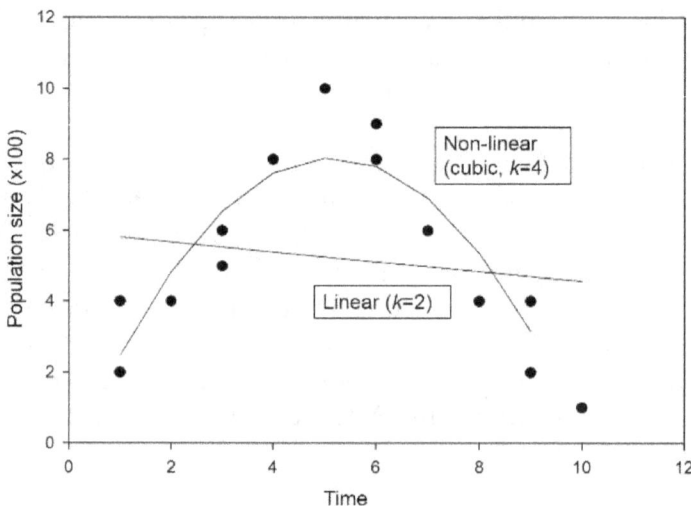

Figure 4.5: *Changes in population size over time (dots) with a comparison of two explanatory models: a simpler linear model and a more complex non-linear model.*

The biologist wants to find the best description of the trend in this data. It is possible that there is only a linear trend through time, but the biologists visually inspects the raw data and sees the possibility that the population has increased and is now decreasing (Figure 4.5).

The linear trend can be quantified with a regression model that estimates an intercept (B_0) and a slope (B_1): $Y = B_0 + B_1(x)$

The non-linear trend could be quantified with a "cubic" model that allows flexibility in the line by adding parameters: $Y = B_0 + B_1(x) + B_2(x^2) + B_3(x^3)$

Thus, we have a simple description of the trend (the linear model, with 2 parameters—the intercept and a slope) that would be easier to estimate and would have fewer variance components. But, it really doesn't look like it 'fits' the data collected by the biologist, does it?

The alternative is a complex model with 4 parameters. It really looks like it "fits" the data much better. But, is there evidence to support the complexity? We can select the best model by assessing the AIC values for each model.

> **AIC does not tell you which model "fits the best"!**
>
> When reporting the results of a model comparison using AIC, <u>avoid the statement</u> *"the model that best fit the data was…"!*
>
> AIC assesses <u>both</u> **fit** <u>and</u> **complexity**. Thus, it is possible that a more complicated model "fits" your data better that the one selected by AIC as the best!
>
> The best way to report the model selected by AIC is: *"The model with the most support, given our data, was…"*

Calculation of AIC values

AIC is calculated with two simple components, which represent the tradeoff between fit and variance:

$$AIC = -2\ln(L) + 2K$$

where:

$L = $ *likelihood for a model under consideration (this describes the "fit" and should be familiar to the reader from our discussion in Chapter 3), and*

$K = $ *number of parameters in the model (this describes the complexity and number of variance components)*

The reader can see that as more parameters are added to the model (as K gets larger), the value for AIC should increase. Hence, the two portions of the equation to obtain the AIC are working against each other.

In the 'game' of using AIC for model selection, the idea is to select the model for which the AIC value is the minimum. If properly applied, the AIC method results in the selection of the best approximating model. That is, of the models under consideration, AIC will select the best approximation of "full reality", or truth, as evidenced by the data that is provided in the sample.

Correction for small sample size?

Sample size can affect the performance of the AIC statistic. Specifically, AIC may perform poorly if there are too many parameters relative to the size of the sample. *We should emphasize that this does not affect the estimate of the parameter (e.g., survival), but it does affect the ranking of the models in competition.*

Thus, many software packages (e.g., SAS proc's, R packages, program MARK) allow the user to examine a version of AIC, corrected for small sample size (**AIC$_c$**):

$$AIC_c = AIC + \frac{2K(K+1)}{n-K-1}$$

One might ask: *"So, how small does my sample need to be to use AIC$_c$ rather than AIC?"*

The answer is—always use AIC$_c$. And, we suggest this not because you should be worried about your sample size in all situations, but rather because AIC$_c$ and AIC become the same as the sample size becomes large. That is, as you increase n in the correction component of the formula for AIC$_c$, the correction component quickly goes to 0.

In Figure 4.6, we show how the AIC correction for a model with 3 parameters ($K=3$) becomes insignificant when the sample goes above 50. So, for samples of more than 50, the value of AIC$_c$ is approximately the same as the value of AIC. If you always use the AIC$_c$ value in your comparison, you won't have to worry about when to switch to AIC...the AIC_c just 'becomes' AIC as the sample increases.

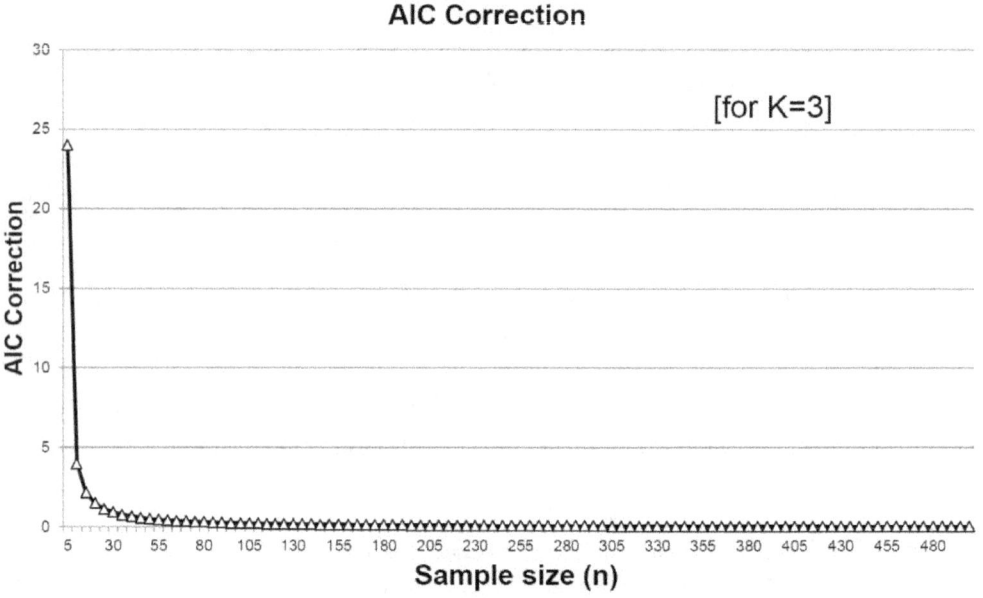

Figure 4.6: *The magnitude of the correction applied to AIC values to create AICc values (corrected for small sample size) as sample size increases from 5 to 500 for a model with 3 parameters.*

Quail example—working through a basic comparison with AIC_c

Let's return to our quail survival example that we have used previously in this chapter. We can stipulate the following models for a simple analysis:

Model	Description	Survival parameters estimated	K
Model 1	All animals have the same probability of monthly survival	$S(.)$	$K = 1$
Model 2	Males and females have distinct survival rates	S_m and S_f	$K = 2$

Let's consider a situation in which we radio-mark 100 adult quail: 50 males and 50 females. All animals are marked and released at the same time, and one month later, the animals are checked using telemetry. We obtain the following data:

20 of the 50 males are still alive
35 of the 50 females are still alive

We are using 'known fate survival' (the topic of Chapter 9), and it is fairly easy to estimate that the monthly survival probability for males is $S_m=0.40$ and $S_f=0.70$ for females. If we pool all animals together, we can see that 55 of the 100 quail survived, so survival of all adults is $S_a=0.55$.

Those rates appear fairly different for males and females, but do we have enough evidence to support that assessment? *Is survival equal for male and female quail?*

To use model comparison with AIC_c to answer our question, we need to splice the information into the equation for the AIC_c, and we can see that we need the values of the model likelihoods (L) and the number of parameters (K).

If you made it through Chapter 3 (Maximum Likelihood Estimation) alive, you should be able to set up this likelihood in your sleep! Let's give it a shot...

For our simple model, Model 1, we need to estimate one survival rate (S_a, adult survival). The likelihood is derived by using our males and our females as two samples of the same statistical population (adults). The males had a sample of 50 "trials", with 20 successes. The females had a sample of 50 "trials" with 35 successes. This can be written as:

$$L(S_a \mid 20,35) = \binom{50}{20} S_a^{20}(1-S_a)^{(30)} \times \binom{50}{35} S_a^{35}(1-S_a)^{(15)}$$

If we work out the math for this (using the estimates of $S_a = 0.55$), we get an estimate for the likelihood,

$$L = 0.0001.$$

Then, we can calculate

$$\ln L = -8.8876.$$

Thus,

$$AIC = -2\ln(L) + 2K$$
$$AIC = -2(-8.8876) + 2(1)$$
$$\mathbf{AIC = 19.7752}$$

Above, we encouraged you to always report AIC_c, with the correction for small sample size, so we need to adjust AIC to AIC_c. We need to know values for K and n; in this model, we have 1 parameter ($K=1$) and our sample size is 100 animals ($n=100$). So, our correction is:

$$AIC_c = AIC + \frac{2K(K+1)}{n-K-1}$$

$$AIC_c = 19.7752 + \frac{2 \cdot 1(1+1)}{100-1-1}$$

$$AIC_c = 19.7752 + \frac{4}{98} = 19.7752 + 0.0408 = 19.8160$$

As expected, with a large sample, the AIC_c (19.81) is very similar to the value for AIC (19.78).

Our more complex model, Model 2, is going to give us two separate survival rates (S_m and S_f: male and female survival). The trials and successes are still the same. We write this likelihood as:

$$L(S_m, S_f \mid 20, 35) = \binom{50}{20} S_m^{20}(1-S_m)^{(30)} \times \binom{50}{35} S_f^{35}(1-S_f)^{(15)}$$

If we work out the math for this (using the estimates of $S_m = 0.4$ and $S_f = 0.7$), we get an estimate for the likelihood,

$$L = 0.0140.$$

Then, we can calculate

$$\ln L = -4.2676.$$

Thus,

$$AIC = -2\ln(L) + 2K$$
$$AIC = -2(-4.2676) + 2(2)$$
$$\mathbf{AIC = 12.5352}$$

Here, we calculate the adjusted AIC_c with $K = 2$ and $n = 100$: the adjustment = 0.1237, and **AIC_c = 12.6589**.

When we compare the two values for AIC_c, we see that the AIC_c for Model 2 is less (AIC_c=12.7) than for Model 1 (AIC = 19.8). Thus, we have evidence from our sample that survival rates do vary for quail between males and females. *Model 2 (sex-specific survival) is closer to reality than is Model 1 (no difference in survival).*

We should note that **our assessment is only as complete as the set of models that we compared**. For example, we did not include age or mass in our assessment. So, our model comparison and the inferences that are derived from it are constrained to the models considered. To investigate a more complete truth about quail survival, we would most likely consider a broader range of models that assess more than a simple gender comparison.

Putting it together: AIC comparison statistics

We can use our newly-calculated AIC values to calculate a set of useful statistics that can help us interpret our results. To what degree is the sex-specific survival model in the bobwhite quail example above, better than the simpler model that suggests survival is the same between sexes?

We can start by calculating **ΔAIC$_c$**, which is the difference between the AIC$_c$ value for any model, i, in our model set and the top-ranked model (Figure 4.2, the model with the minimum AIC$_c$ value, or AIC$_{c\_0}$):

$$\Delta AIC_c = AIC_{c\_i} - AIC_{c\_0}$$

To interpret the relative support for a given model we need to calibrate the assessment of that model relative to the entire set of models under consideration. We use the ΔAIC to calculate **AIC weights** (w_i) to provide an index of this evidence—the likelihood of a model, given the model set. We calculate the model weight (w_i) for the ith model in the set of n models as:

$$w_i = \frac{\exp\left(-\frac{1}{2} \cdot \Delta AIC_{ci}\right)}{\sum_{i=1}^{n} \exp\left(-\frac{1}{2} \cdot \Delta AIC_{ci}\right)}$$

The next statistic that we may calculate uses the model weights to express the **model likelihood** (ML, or, literally, "how likely") for a given model. The model likelihood is the ratio of the model weight for model i, compared to the top-ranked model's weight (w_0):

$$ML_i = \frac{w_i}{w_0}$$

We interpret the model likelihood value as the strength of evidence of this model, relative to other models under comparison.

We can place these statistics, calculated for our bobwhite quail example, in a model comparison table that would be suitable for use in a publication (where k is the number of parameters for model i):

Model	AIC$_c$	ΔAIC$_c$	w_i	Model likelihood	k
S(sex-specific)	12.659	0	0.973	1.000	2
S(.)	19.826	7.167	0.027	0.028	1

Now, we can interpret our results, fully. The top-ranked model has approximately 97% of the model weight in our simple model set of two models. The second-ranked model only has 3% of the model weight, and the top-ranked model is 36 times more likely to be the best model than the second-ranked model (1/0.028=36).

We recommend using **"evidence-based language"** when describing the results of model comparison. Here, there is "strong evidence" for the top-ranked model. If model weights were more similar, our evidence would start to fade. We might use phrases like "good evidence", "some evidence", "little evidence", or "no evidence" to describe our certainty that males and females have different survival rates, based on model results.

As a rough guideline, Burnham and Anderson (2002, p. 70) provide the following recommendations for interpretation of AIC-based model comparisons:

ΔAIC_i	Level of empirical support for model i
0–2	"Substantial"
4–7	"Considerably less"
> 10	"Essentially none"

Thus, our example of bobwhite quail survival is an easy interpretation—we have strong evidence as the support for the null model, $S(.)$, is "considerably less" than support for the sex-specific model.. But, what if our results were different? What if we had weaker evidence that the sex-specific model was the best? What if there was "substantial" support for the null model, $S(.)$? Let us consider a *different outcome* from this model comparison, to explore some details of model selection:

Model	AIC_c	ΔAIC_c	w_i	Model likelihood	k
S(sex-specific)	12.659	0	0.630	1.000	2
S(.)	13.600	1.065	0.370	0.587	1

Now, the weights of the two models are more similar. The ΔAIC for the second-ranked model is <2.0, and the table above suggests that our second-ranked model would have "substantial support", even though it was ranked second. And, our model likelihood suggests that our top-ranked model is not even twice as likely to be the best model as our second-ranked model (1.000/0.587=1.7). So, it is very difficult for us to distinguish between our two models.

In such a situation, model selection theory provides an approach that we could use to select the best model—the concept of **parsimony.** *Parsimony suggests that, given equal explanatory value, we should select the <u>simplest</u> explanation.* We measure simplicity as the number of parameters, so the $S(.)$ model (with 1 survival rate and 1 parameter) would be selected as the best model. Another way to view this situation is that the sex-specific survival model had not accumulated enough evidence to distance itself from the simpler model, and thus we have very little evidence to suggest that quail have sex-specific survival rates.

Part Two: Thoughts on model selection (when life gets messy)

In fact, model comparison/selection can become very difficult when one model is not a clear 'best model'. Should we consider all models plausible if they are within 2 ΔAIC from the best model? It is not unusual for ecologists to throw up their hands in frustration when they receive results such as those shown in the table below:

Model	AIC$_c$	ΔAIC$_c$	w_i	Model likelihood	k	-2ln(L)
Model 1	187.30	0	0.40	1.000	2	183.3
Model 2	188.50	1.2	0.22	0.549	3	182.5
Model 3	188.60	1.3	0.21	0.522	2	184.6
Model 4	189.10	1.8	0.16	0.407	6	177.1

However, we can find a way forward when we find ourselves in this situation. Let's review what we know about our results, keeping in mind the formula for AIC:

$$\text{AIC} = -2\ln(L) + 2K$$

- *Model 1:* was ranked the top model, and it is a simple model with only 2 parameters.
- *Model 2:* we can see that this model's 'fit' (-2lnL) was actually a bit better (lower value) than Model 1. But, the AIC formula penalizes (with the term $2*K$) this model by 2 AIC for the one additional parameter ($K=3$), compared to Model 1. Thus, Model 2 fits a little better than Model 1, but not enough to make up for the complexity.
- *Model 3:* this model's 'fit' is not as good (larger value for -2lnL) as Model 1, and it has exactly the same number of parameters as Model 1 ($K=2$).
- *Model 4:* this model's 'fit' is also better (smaller value for -2lnL) than Model 1. But, AIC penalizes Model 4 a whopping +8 for four additional parameters ($K=6$), relative to Model 1. The better fit is not enough to make up for the complexity of the model.

Neither Model 2 nor Model 4 have improved their likelihood *enough* to counter the addition of additional parameters. And, although Model 4 has a better fit (as measured by (-2lnL), it is not sufficiently better to merit the addition of 4 more parameters relative to Model 1. As such, we can suggest that the additional parameters in Model 2 and Model 4 are *uninformative parameters*. The additional parameters are not helpful to describe reality, as judged by the evidence we have collected (our data).

With reference to our assessment of Model 2, Burnham and Anderson (2002, p. 131) wrote:
> "Models having Δ_i (ΔAIC) within about 0-2 units of the best model should be examined to see whether they differ from the best model by 1 parameter and have essentially the same values of the maximized log-likelihood as the best model. In this case, the larger model is not really supported or competitive, but rather is 'close' only because it adds 1 parameter and therefore will be within 2 Δ_i units, even though the fit, as measured by the log-likelihood is not improved."

Arnold (2010) provided a useful summary of ways forward for ecologists when we encounter this situation of models that appear to have uninformative parameters (many models within 2 ΔAIC of the top model). Arnold (2010) evaluated the merits of five options:

- Full reporting when model sets are small
- Model averaging
- Assessment of confidence intervals to evaluate evidence
- Assessment of relative variable importance
- Discarding models with uninformative parameters

In the next few sections of this chapter, we will follow Arnold's (2010) thoughts with some examples of our own.

Normally, the practice of **full reporting** is possible only when the model set is small (perhaps fewer than 10 models). In this situation, we would create a table that shows all 10 models, and we would report on the levels of uncertainty (e.g., confidence intervals for parameter estimates) and we would evaluate the strength of the evidence. We could interpret the AIC values for each model in light of the number of parameters. We might use the concept of **parsimony**, for example, to do a complete assessment of our models with a mind to use the simplest explanation of reality if evidence does not accumulate for more complex examples. An example of full reporting is the detailed evaluation of Models 1, 2, 3, and 4 that we provide above.

Ecologists may also use a process known as **model averaging** to incorporate **model uncertainty** into the estimates of parameters. This approach focuses on the estimate of parameters obtained from individual models in the data set and uses a weighted average across the estimates from more than one model. In our quail example, to obtain a model-averaged value for males in the population, we could use a weighted average (weighted by the model weights, w_i) of the male survival rate from the sex-specific model and the pooled survival that would be attributed to males (and also females) in the S(.) model. *However, we note that model averaging is not recommended in our first set of quail results, because we are highly certain (based on model weights) that survival is sex-specific.*

The approaches to model averaging are numerous, and the reader is encouraged to read more about model averaging before conducting the process. The first decision in model-averaging is to *decide which of your models to include in the average*. Some ecologists, for example, model average across the set of models that are within 7 ΔAIC of the top model. Other ecologists select the model set for averaging that, as a group, make up 90-95% of the cumulative model weight—in our example above, we would use all four models by either set of rules. Rehme et al. (2011) provide a review of the complex set of approaches that have been used by biologists to establish a model set for model averaging.

However, we agree with Arnold (2010) when he states that if models with uninformative parameters were removed prior to model averaging, the top model might often have 80-90% of the model weight, and we would often have no need to model average. The argument to avoid model averaging when possible is that the SE calculated for our model-averaged parameter is always larger than the standard SE estimated using maximum likelihood estimation for the parameter in an individual model. The SE becomes larger, because we incorporate **model**

uncertainty into the SE when we model average—and thus, we incorporate more error. Although this does account for uncertainty in the search for truth, we'd like to avoid inflating variance estimates when we can—especially if the manner in which we constructed our models added to the confusion. However, in the spirit of presenting all sides of the argument (we told you this could be complicated, right?!), we note that other quantitative scientists are firmly supportive of model averaging (e.g., Burnham and Anderson 2002, Lukacs et al. 2010, Doherty et al. 2012).

Two methods for **model averaging** are used when model certainty is low. **Unconditional model averaging** is used to calculate a weighted estimate by considering all models, R, in the model set, regardless of whether they contain the covariate of interest, β_j.

$$\bar{\beta}_j = \sum_{n=1}^{R} w_i \hat{\beta}_{j,i}$$

Conditional model averaging also calculates a weighted estimate for a covariate, but the average is *conditioned* on whether the covariate, β_j, appears in the model—only such models and their summed weights (denominator, below) are used for averaging.

$$\bar{\beta}_j = \frac{\sum_{n=1}^{R} w_i \hat{\beta}_{j,i}}{\sum_{n=1}^{R} w_i}$$

Let us consider an assessment of survival that includes 5 models. Each model contains covariates that describe potential effects of time (t), vegetative cover, and age of the individual on survival:

Model	β_{cover}	w_i	Unconditional	Conditional
1: S(cover, t)	2.16	0.54	X	X
2: S(t)	-	0.35	X	
3: S(cover)	2.57	0.05	X	X
4: S(age)	-	0.04	X	
5: S(age, t)	-	0.02	X	

If we decide to obtain a model-averaged value for the cover covariate, β_{cover}, we have two choices. The table shows the value for the covariate of cover (β_{cover}) for the two models in which the covariate is present (Model 1: β_{cover} =2.16; Model 3: β_{cover} 2.57). Unconditional model averaging will use the model weights and the values for β_{cover} to average across all five models in our model set. **Note that when "cover" is not present in the model, unconditional model averaging uses β_{cover} =0.**

The **unconditional** model averaged estimate:

$$\overline{\beta}_{cover} = (2.16*0.54) + 0 + (2.57*0.05) + 0 + 0$$
$$= 1.17 + 0 + 0.13 + 0 + 0$$
$$= 1.30$$

Alternatively, a **conditional** model averaging approach would only use the models for which β_{cover} is included, and the sum of the weights for those models:

$$\overline{\beta}_{cover} = \frac{(2.16*0.54) + (2.57*0.05)}{0.54 + 0.05}$$
$$= \frac{1.17 + 0.13}{0.59}$$
$$= 1.30 / 0.59$$
$$= 2.20$$

It may seem odd to use $\beta_{cover} = 0$ for models in which β_{cover} does not exist. But, proponents of unconditional model averaging suggest that if a model such as S(age) hypothesizes that age has an effect on survival, we also have an unwritten hypothesis: *the S(age) model suggests that cover does NOT affect survival*. Thus, $\beta_{cover} = 0$ by default in the S(age) model.

As your authors, we will admit a personal bias against unconditional model averaging. It seems fairly obtuse, to us, to go to the trouble of finding an unbiased, maximum likelihood estimate for a covariate value, only to modify (*bias?!*) that estimate through an averaging process that could vary tremendously because of the number of models and the structure of the models submitted for consideration. *We tend to favor conditional model averaging, if model averaging is necessary.* **Above all, we encourage you to consider your analysis (will you model average?) before you construct your model set.**

Building model sets: plan ahead

As an example of how the structure of the set of models submitted in an analysis can affect the results when model averaging, let us consider the following two sets of models:

Model Set 1:
S(age)
S(t)
S(gender)
S(.)

Model Set 2:
S(age)
S(t)
S(gender)
S(.)
S(age+t)
S(age+gender)
S(gender+t)

If we were interested in a model-averaged estimate of the covariate for "gender" in this analysis, which model set is better planned for the use of model averaging? We would submit that the second set is best constructed for averaging. Doherty et al. (2012) also suggest 'balanced' sets of models for use in model averaging.

In fact, it would seem that a model averaged estimate for 'gender' is nonsensical in Model Set 1, as there is only one model that includes gender. By default, there is no other model to average across—and if unconditional model averaging is used, the estimate will be affected by the $\beta_{gender} = 0$ found in the other three models.

In Model Set 2, "gender" appears in three models—we could average across three things, logically.

We will also note that the first set of models has the underlying hypothesis that only age OR time OR gender affect survival, but not more than one effect. And, if both age and gender have an effect on survival (a common occurrence in nature), both the age and gender model should be highly supported…but neither will separate itself as the 'best model'. In truth, there are two good, single-factor models [S(age) and S(gender)]. Such a dynamic would be documented, most likely, by the use of the second set of models, which includes models that propose two influences on survival [e.g., S(age+gender)].

AIC values are only one portion of your results: don't forget to look at your estimates!

Thus far, we have only looked at the **model rankings** to document the strength of evidence found in our analyses. But, of course we can also look within each model to assess the parameter estimates and the variance—often expressed as a **95% confidence interval**. If the confidence intervals for male and female survival overlap, that is good evidence that males and females do not have distinct survival rates. However, if confidence intervals do NOT overlap, it is good evidence to suggest that males and females do have distinct survival rates.

However, Arnold (2010) pointed out that ecologists may find evidence that appears to conflict when they compare AIC rankings with information from 95% confidence intervals of parameter estimates. For example—in some instances, the AIC rankings may indicate that a sex-specific model (or other type of more complex model) is ranked higher than a null model with no effect. The ΔAIC for the null model may be >2.0, leading us to believe that we should favor the sex-specific model. But, when we look at the 95% confidence intervals for male and female survival estimates, the intervals overlap—suggesting that there is not an effect of sex on survival. Thus, the AIC comparison and the 95% confidence interval are sending conflicting messages. If this hasn't happened to you yet in your career, it will happen to you at some point!

We encourage the reader to evaluate Arnold's (2010) thoughtful analysis of this situation in detail, but we can summarize the explanation. We know from the information early in this chapter that AIC provides a +2 penalty for adding an additional parameter. When we use ΔAIC>2.0 as our guide to tell us when one model is better than another, this is equivalent to using an α-level of 0.15, rather than $\alpha = 0.05$.

As we know, the 95% confidence interval is based on α =0.05. So, it does not make sense to evaluate parameters at the 95% confidence interval level—we should, in fact, use an 85% confidence interval for our comparison of male and female survival rates. An 85% confidence interval is "tighter" (narrower) than a 95% confidence interval. And, this explains why we sometimes see AIC comparisons that suggest survival rates are different between groups, yet the 95% CI's overlap for the groups.

If you use software for your analysis that allows you to set the confidence intervals provided, we suggest that you follow Arnold's (2010) advice to create 85% confidence intervals. However, some software packages (e.g., PRESENCE, MARK) do not allow this flexibility as of this writing.

Cumulative weight: more evidence

Another strategy to assess the value of a covariate in a more complex model analysis, especially when model certainty is low, is to analyze the **relative importance** of each variable. In this case we determine the cumulative weight of all models in which the parameter is found—the logic is that if a variable is important, most of the models in which it appears should have high model weights, and thus the cumulative weight for that variable will be high.

As Burnham and Anderson (2002) suggest, we must take care to not create implausible hypotheses and miss the point of "thoughtful" model sets. Thus, we must think about how we construct models for comparison. Specifically, model sets must be constructed in "balanced" fashion—with balanced combinations of variables that have biological meaning.

Model	Weight
Grass	0.35**
Day + Grass	0.25**
Gender	0.12
Hormone	0.10
Grass + Gender	0.08**
Null	0.05
Grass + Forb	0.05**

In the model set above, the cumulative weight for the effect of grass cover = 0.73 (the sum of all weights marked with "**"). Is grass cover an important variable? We hardly know—because grass appeared in all but three models. The effect of day and forb cover only appear in one model. That's not a fair comparison, no matter how you think about it.

Model	Weight
Grass	0.35**
Day + Grass	0.25**
Null	0.12
Hormone	0.10
Day	0.08
Hormone + Day	0.05
Hormone + Grass	0.05**

The second set of models, above, is balanced with one-factor and two-factor models. Each variable, above, appears in three models. We find that the cumulative weights for grass = 0.65 (i.e., 0.35+0.25+0.05 = 0.65), while cumulative weight for day = 0.38 and hormone = 0.20. Thus, we can conclude—much more legitimately—that grass is an important factor to describe variation in the parameter of interest.

Another approach is to avoid situations with high model uncertainty through careful planning and the use of **step-down model selection**. We can use a step-by-step process to eliminate models that are not useful, which will limit the number of models under consideration in the final model comparison. Lebreton et al. (1992) described this process as "step-down" analysis, and they recommended it in situations when model sets were prone to be unmanageably large if an "all possible combinations" approach was taken (see Chapter 13 for an example of a robust design analysis that evaluated models with 150-500 parameters). In fact, program Distance (Buckland et al. 2001; Chapter 19) uses a similar approach to step through the selection of the best model to describe detection patterns in surveys.

If your analysis has a small number of parameters and associated covariates to estimate, it seems reasonable to follow Doherty et al. (2012) who recommended an "all combinations" model strategy. Such an approach provides balanced sets of models that can be used in model averaging and it avoids ad hoc strategies of model selection. In fact, the simulation study by Doherty et al. (2012) suggested that ad hoc strategies ran the risk of inflating the importance of variables that were weakly correlated with the parameters of interest.

Here is an example to explain the "step-down" model selection approach that can simplify your model set. Let us assume that we have a goal to assess the effects of age, sex, vegetation density (of the current location) on the survival of the northern bobwhites we used in a previous example. Sex and age are categorical variables (male/female and juvenile/adult), but, vegetation density is a continuous variable. We might postulate that vegetation density has a linear effect on survival (as vegetation density increases, survival increases). However, we also might postulate that vegetation density has a non-linear effect on survival—perhaps survival is low when quail use sparsely vegetated areas and also low when the quail use very densely vegetated areas.

Thus, we develop two models that describe the possible effect of vegetation density on survival:

Linear: $S = B_0 + B_1(\text{veg. density})$
Non-linear: $S = B_0 + B_1(\text{veg. density}) + B_2(\text{veg. density}^2)$

To use step-down model comparison, we would first use AIC to compare our two vegetation density models to see which best describes variation in survival. This step is exactly like the process used by program Distance to select the shape of detection functions for surveys (Buckland et al. 2001; see Chapter 19).

Once a model is selected (e.g., linear), that model is put forward with the other models for a final comparison:

Model 1 (sex): $S = B_0 + B_1(\text{male})$
Model 2 (age): $S = B_0 + B_1(\text{adult})$
Model 3 (vegetation): $S = B_0 + B_1(\text{veg. density})$
Model 4 (null, no effect): $S = B_0$

In this simple model set, our final list of models is only reduced by one model through step-down model selection. But, you can imagine a more complicated scenario with multiple covariates that might benefit from assessing linear and non-linear effects in a first step, before the final selection process. In Chapter 19, we will discuss program Distance (Thomas et al. 2010), and this software uses a 3-step, step-down model selection process to find the best model to describe the decline in detection probability with distance from a transect or point.

To conclude our suggestions for model comparison under low levels of certainty, we suggest (*following* Arnold 2010):
- <u>If you have small sets of models:</u> use full reporting, avoid model averaging "at all costs", and provide interpretation with 85% confidence intervals and assessment of the number of parameters and relative model fit (likelihood).
- <u>If you have model sets with many variables:</u> use balanced sets of models, and report the relative importance using cumulative model weights. Use sequential, or step-down modeling to remove uninformative parameters and/or models.
- <u>For all circumstances:</u> try to avoid model averaging if at all possible. If model averaging must be used, construct your model set to prepare for the potential use of model averaging. *Do not make model averaging a last-minute thought!* And, construct models sets that are conducive to proper application of model averaging. Consider conditional model averaging to avoid the loss of information from estimates derived from unbiased maximum likelihood estimation methods.

Conclusion

AIC is a method that allows comparison of the support provided to multiple models by our data. As the investigative biologist, we want a good, descriptive model that informs us about our population. But, we also value simplicity, because simplicity allows us to limit the variance components in a model. Parsimony suggests that, given equal explanatory value (fit, as measured by likelihood), we select the simplest explanation. AIC evaluates the trade-off between model fit and variance components to suggest which model is closest to reality. Model selection is, conceptually, a simple idea; but the realities of model selection can be very messy! We encourage you to evaluate the various information provided by your analyses (model rankings and weights, presence of uninformative parameters, cumulative model weights, and estimates of coefficients from models) to make sense of model selection and parameter estimation. Above all, plan ahead when constructing models for analysis.

References

Anderson, D. R. 2007. Model based inference in the life sciences: a primer on evidence. Springer Science & Business Media.

Arnold, T. W. 2010. Uninformative parameters and model selection using Akaike's Information Criterion. The Journal of Wildlife Management 74: 1175-1178.

Buckland, S. T., D. R. Anderson, K. P. Burnham, J. L. Laake, D. L. Borchers, and L. Thomas. 2001. Introduction to Distance Sampling: Estimating Abundance of Biological Populations. Oxford University Press, New York.

Burnham, K. P., and D. R. Anderson. 2002. Model Selection and Multimodel Inference. Springer-Verlag, New York

Chamberlin, T. C. 1965. The method of multiple working hypotheses. Science 148(3671): 754-759.

Doherty, P. F., G. C. White, and K. P. Burnham. 2012. Comparison of model building and selection strategies. Journal of Ornithology 152: 317-323.

Hines, J.E., and J.R. Sauer. 1989. Program CONTRAST: A General Program for the Analysis of Several Survival or Recovery Rate Estimates. US Fish & Wildlife Service, Fish & Wildlife Technical Report 24, Washington, DC.

Lebreton, J. D., K. P. Burnham, J. Clobert, and D. R. Anderson. 1992. Modeling survival and testing biological hypotheses using marked animals: a unified approach with case studies. Ecological Monographs 62: 67-118.

Lukacs, P. M., K. P. Burnham, and D. R. Anderson. 2010. Model selection bias and Freedman's paradox. Annals of the Institute of Statistical Mathematics 62: 117-125.

Rehme, S. E., L. A. Powell, and C. R. Allen. 2011. Multimodel inference and adaptive management. Journal of Environmental Management 92: 1360-1364.

Thomas, L., S. T. Buckland, E. A. Rexstad, J. L. Laake, S. Strindberg, S. L. Hedley, J. R. B. Bishop, T. A. Marques, and K. P. Burnham. 2010. Distance software: design and analysis of distance sampling surveys for estimating population size. The Journal of Applied Ecology, 47: 5–14.

For more information on topics in this chapter

Conroy, M. J., and J. P. Carroll. 2009. Quantitative Conservation of Vertebrates. Wiley-Blackwell: Sussex, UK.

Cooch, E., and G. White. 2014. Chapter 4: Building and Comparing Models. *In* Program MARK: a gentle introduction, 12th edition, Cooch, E. and G. White, eds. Online: http://www.phidot.org/software/mark/docs/book/pdf/chap4.pdf

Williams, B. K., J. D. Nichols, and M. J. Conroy. 2002. Analysis and management of animal populations. Academic Press, San Diego.

Citing this primer

Powell, L. A., and G. A. Gale. 2015. Estimation of Parameters for Animal Populations: a primer for the rest of us. Caught Napping Publications: Lincoln, NE.

A biologist prepares to release a channel catfish (Ictalurus punctatus) with individually numbered t-bar tags (Floy type) near its dorsal fin on the Red River near Selkirk, Manitoba, Canada. Photo provided by Stephen Siddons, University of Nebraska-Lincoln (used with permission).

Ch. 5

Variance and the delta method[1]

"The recipe for life is not that complicated. There are a limited number of elements inside your body. Most of your mass is carbon, oxygen, hydrogen, sulfur, plus some nitrogen and phosphorous. There are a couple dozen other elements that are in there in trace amounts, but to a first approximation, you're just a bag of carbon, oxygen, and hydrogen."
-- Andrew H. Knoll

Questions to ponder:

- *I have a variance estimate for daily survival—how do I get a variance estimate for annual survival after I raise survival to the 365th power?*
- *What steps do I follow to implement the delta method?*
- *How do I know how to find a derivative?*
- *Seriously, you want me to find a derivative?*

Into the real world: a need for variance

Biologists who estimate the value of parameters, such as survival, density, or capture probability, also need an estimate of the variance of these parameters. We might use the variance to construct a 95% confidence interval as we compare the survival from our study to the probability of survival reported by another study, for example. In Chapter 2, we discussed the concept of precision, and the magnitude of confidence intervals around an estimate is important (Figure. 2.5). Also, in Chapter 3, we discussed how maximum likelihood estimation allows us to obtain estimates of variance.

We can look at a hypothethical example to see why estimation of variance and establishment of confidence intervals is important. Let us assume that we have conducted a radio-telemetry study (Chapter 9) of two populations of animals, and we have estimated the annual survival probability

---

[1] *With thanks for content to Christopher Chizinsky, Michael Conroy, Evan Cooch, and Gary White*

(Figure 5.1). Population A lives in a heavily agricultural area and Population B lives in an area less affected by human land use. *Do the two populations' survival estimates differ?*

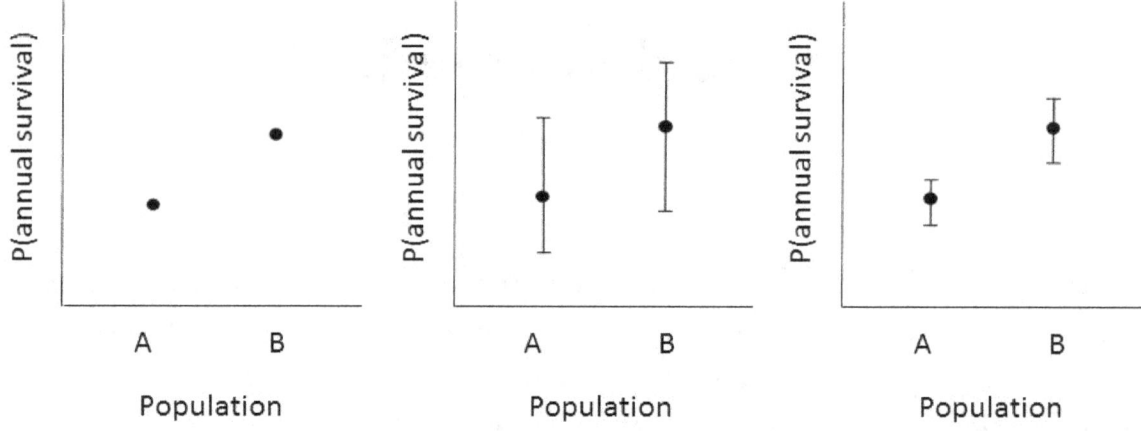

Figure 5.1: *Comparison of information provided for estimates of annual survival with (center, right) and without (left) estimates of precision with 95% confidence intervals. The inference drawn from each scenario is also different. At left, differences in survival between study sites cannot be assessed without information about precision of the estimates. Center, the two populations do not have different annual survival (confidence intervals overlap). At right, evidence suggests the annual survival of the two populations is different (confidence intervals do not overlap).*

Figure 5.1 shows (at left) what would be lacking if we did not provide any estimate of the precision of our survival estimates. The survival probabilities of Population A and B *look* different. But, are they? If our study had relatively low precision, our results would look like the center portion of Figure 5.1—large, overlapping confidence intervals. Our data are not able to show any differences—survival is similar in the two populations. But, if our study had higher levels of precision, we might get a result like that shown in the right portion of Figure 5.1. Here, the confidence intervals are relatively smaller and, importantly, they do not overlap—we have evidence that the probability of annual survival is different in the two populations.

Thus, estimates of variance—which can be converted to, and reported as, standard errors and 95% confidence intervals—are extremely important to the inferences of our studies. *We cannot go to all of the work of estimating our parameter's value without providing the information to interpret the level of certainty we have in our estimate.*

Luckily, most software packages (those which you will use to estimate survival, abundance, density, population size, movement probabilities, or other parameters) also provide estimates for standard error or the 95% confidence interval for your parameter estimate. But, variance estimates become problematic when biologists are required to change temporal scales for a parameter—perhaps you have estimated daily probability of nest survival, and you need to extrapolate the daily survival to estimate the 24-day "complete" nest survival estimate? We also encounter problems with variance when we need to combine demographic parameter estimates to indirectly calculate a demographic parameter—perhaps you have estimated "complete" nest survival ($\hat{S}$) and you would like to estimate fecundity (average number of offspring produced, F) as $\hat{F} = \hat{S} *$ clutch size. Both $\hat{S}$ and clutch size have estimates of variance, but how do you obtain a variance for $\hat{F}$? Or, perhaps we want to calculate the mean of a set of demographic parameters across years—for example, if we had annual estimates for density ($\hat{D}$) for a three-year study. How do we get the variance of the mean $\hat{D}$? The topics discussed in this chapter become important in any of these situations.

In short, our problem arises any time we want to "do something" with our estimate of a parameter that involves placing it into a formula to calculate something else. And, this happens quite often—in fact, the reason we estimate parameters is typically to "do something" with them, correct?!

In each of the three preceding examples, the new demographic parameter ("complete nest survival", fecundity, or mean density) is a function of at least one other variable. Thus, it makes logical sense that the variance of the new parameter is also a function of the variance of the former parameters. But, the estimation process for the new variance is not straightforward. For example, if we estimate complete nest survival (24-day survival, S_{24}, for a songbird like the wood thrush [*Hylocichla mustelina*]) as $S_{24} = S_d^{24}$, we <u>cannot</u> simply follow the same process for the variance, therefore $\text{var}(S_{24}) \neq \text{var}(S_d)^{24}$.

> ### Sampling variance
>
> **Sampling variance** is the variance attributed to estimation of a parameter from sample data—affected by variation among individuals and the sample size. Alternatively, **process variance** describes variation in a parameter over time and space—affected by long-term variation in weather, disease, and other biological forces.
>
> *The standard error (SE) and 95% confidence interval provided by your analysis software (MARK, R, SAS, etc.) is an estimate of sampling variance.* The discussion in this chapter is applicable to sampling variance. If you are interested in simulation modeling, you should consider methods to estimate or approximate or account for process variance (over space and longer time periods).
>
> **Good sources** for discussion about process variance: White et al. (1982), Franklin et al. (2000).

Delta method to the rescue

In this situation, many smart folks have suggested the use of the delta method (Seber 1982, Hilborn and Mangel 1997:58–59, Williams et al. 2002:736, Skalski et al. 2005:570–571, Cooch and White 2014, MacKenzie et al. 2006:66, 73–75). So, there appears to be a solution.

However, many well-meaning biologists have gone to these texts for help and met a bewildering set of formulas that are not particularly useful to the first-time (or even more advanced!) user. Feeling befuddled by equations, biologists have continued to publish critical comparisons of demographic parameters without estimates of variance to guide decision-making or hypothesis evaluation. We'd like to change that—we hope the coverage here is useful to the first-time user.

Before we launch into a discussion of the method, we'd like to note that, as the old saying goes, "There is more than one way to skin a cat." The delta method works best for functions that are largely linear in nature, and the variance of complicated functions may not be approximated well with the delta method. For example, population or ecosystem models with many variables may be better served with another method. Some scientists use a method known as **bootstrapping**, which makes use of simulations of a stochastic model. After evaluating the results of perhaps 1000 simulations, a 95% confidence interval can be assessed by removing the upper 2.5% of estimates and the lower 2.5% of model-predicted values.

So, bootstrapping can be a very useful approach for more complex models. Regardless, the delta method has plenty of uses under simpler situations in ecology, as we will see.

The delta method has been around for a long time in the world of statistics (see a review by Ver Hoef 2012). The delta method approximates variance in complex situations where variance cannot be easily derived. Generally, we will use a method from calculus, a *Taylor series expansion*, to derive a linear function that approximates a more complicated function.

Two parts of this explanation are important for us: (1) the variance that we derive from the delta method is an **approximation** of the variance, **not an estimate** of the variance, and (2) we are so happy to see our friend, calculus, again (refer back to Chapter 3, maximum likelihood estimators for more fun with calculus)!

The basic formula used to approximate the variance of a function, G, for which the variance is a function of one or more random variables (these will be variables of interest), X_1 to X_n, is as follows:

$$\text{var}(G) = \text{var}[f(X_1, X_2, ..., X_n)] = \sum_{i=1}^{n} \text{var}(X_i) \left[\frac{\partial f}{\partial X_i} \right]^2$$

We can use applications of this general formula to explore using the delta method for a situation in which our new variance is a function of only one random variable (X_1) and its variance—such as our "daily to 24-day" nest survival example ($X_1 = S_{daily}$):

$$\text{var}(G) = \text{var}[f(X_1)] = \text{var}(X_1) \left[\frac{\partial f}{\partial X_1} \right]^2$$

We can also use the formula for the delta method for a situation in which our new variance is a function of two random variables (X_1 and X_2) and their variance, such as the example of fecundity used earlier, a function of nest survival (X_1) and clutch size (X_2):

$$\text{var}(G) = \text{var}[f(X_1, X_2)] = \text{var}(X_1) \left[\frac{\partial f}{\partial X_1} \right]^2 + \text{var}(X_2) \left[\frac{\partial f}{\partial X_2} \right]^2$$

Derivatives

You will notice the portion of the equation for the delta method that involves a derivative of a function, f, with respect to a specific random variable, X_1:

$$\left[\frac{\partial f}{\partial X_1} \right]$$

We need to remind ourselves how to calculate derivatives, as that is a critical part of the process to obtain the approximation for our new variance. We can open any calculus textbook and find a table similar to the following (after Powell 2007):

Function, generalized	Generalized derivative	Function, example	Derivative, example
c	0	2	0
x	1	S	1
cx	c	$2S$	2
x + c	x	$S + 2$	S
x^c	$cx^{(c-1)}$	S^2	$2S^1$
$\dfrac{x}{c}$	$\dfrac{1}{c}$	$\dfrac{S}{2}$	$\dfrac{1}{2}$
$\dfrac{c}{x}$	$-\dfrac{c}{x^2}$	$\dfrac{2}{S}$	$-\dfrac{2}{S^2}$

Does this bring back any memories from calculus class?!

Step-by-step: working some examples

Let's start with a simple example with one variable involved. Suppose you have estimated an annual probability of survival, and its associated variance component (likely reported as a standard error, SE, by the software). Impressed with your study and the results, you write a manuscript and send it to the Journal of Amazing Results. The reviewers love your paper, and the Editor loves your paper. However, the Editor requests one "small change"—she wants you to report your survival rate as a mortality rate, instead.

"That's easy," you respond and you proceed to make the change.

Figure 5.2: *The probability of survival and the probability of mortality for wood thrushes (*Hylocichla mustelina*) are the complement of one another. Photos by L. Powell.*

It is easy to provide the estimate for mortality rate (Figure 5.2):

$$\hat{M} = 1 - \hat{S}$$

And, then you stop—how are you going to obtain the variance (or the SE) for $\hat{M}$?

Let's use the delta method to approximate the variance for $\hat{M}$. There are **4 simple steps**:

1—write down the equation (the function) of what you want to estimate. We want to estimate mortality rate, and we've already written that function as:

$$\hat{M} = 1 - \hat{S}$$

2—write out the variance of the new parameter in "delta method equation form". Our new parameter, M, is the function of only one random variable, S. The equation to approximate the variance of M can be written as:

$$\text{vâr}(\hat{M}) = \text{vâr}(\hat{S}) \left[\frac{\partial \hat{M}}{\partial \hat{S}} \right]^2$$

3—find the derivatives you need for your function. The derivative of the function M, with respect to the random variable, S, can be obtained as follows:

$$(1 - S)' = -1$$

We find the derivative of 1-S by separating it—we can find the derivative of [1] and the derivative of [–S] and add those derivatives together. The derivative of a constant (c), such as 1 in our function, is 0. Therefore, this will disappear because it equals 0. And the derivative of –S can be visualized as the derivative of (-1)(S). We can find in our table above that the derivative of the product of a constant and a random variable (*cx*) is x. So, the derivative of (-1)(S) is -1. We add the two sub-derivatives together: 0 + (-1) = -1.

4—insert the derivative and arrive at the equation we need for our variance approximation. Here, we have:

$$\text{vâr}(\hat{M}) = \text{vâr}(\hat{S}) \left[\frac{\partial \hat{M}}{\partial \hat{S}} \right]^2 = \text{vâr}(\hat{S})(-1)^2$$

When we simplify by squaring the (-1), we notice that:

$$\text{vâr}(\hat{M}) = \text{vâr}(\hat{S})$$

Thus, the variance of $\hat{M}$ is actually the same value as the variance for $\hat{S}$! That makes sense, when we think about it—we are just as uncertain about the mortality rate as we were the survival rate, and the two probabilities are 'on the same scale'—we are just expressing S as its complement, M.

Here's another example with only one random variable. Suppose that you had estimated the density of a grassland bird as:

$$\hat{D} = 15.3 \text{ birds/ha.} \quad SE = 1.1$$

During the publication process, a journal editor asks you to report your density estimate in a standard unit for avian ecologists—the number of birds per 10 ha. You can easily determine that the new density is $\hat{D}_{10ha}$ 153 birds/10 ha (*calculated as:* 15.3*10 = 153).

But, we need the associated variance. And, this one is trickier than the survival/mortality estimate that we just completed. Now that you're familiar with the delta method, let's work through it…

Figure 5.3: *Density of grassland birds can be estimated with visual surveys (Photo by S. Kempema)*

Step 1: write down the equation (function) of what you want to estimate:

$$\hat{D}_{10ha} = 10 \cdot \hat{D}_{1ha}$$

Step 2: write out the variance of the new parameter in delta method equation form (we have only one random variable, so it's an easy application):

$$\text{vâr}(\hat{D}_{10ha}) = \text{vâr}(\hat{D}_{1ha}) \left[\frac{\partial \hat{D}_{10ha}}{\partial \hat{D}_{1ha}} \right]^2$$

Step 3: find the derivatives you need for your function (*Note: our derivative matches the cx in our table*):

$$(10\hat{D}_{1ha})' = 10$$

Step 4: arrive at the equation you need by inserting the derivative into the formula in Step 2:

$$\text{vâr}(D_{10ha}) = \text{vâr}(\hat{D}_{1ha})10^2$$

And, simplifying:

$$\text{vâr}(D_{10ha}) = 100\,\text{vâr}(\hat{D}_{1ha})$$

Now that we have our equation, we can calculate our new approximated SE. The relationship between SE and variance is:

$$SE = \sqrt{\text{var}}$$

To calculate var(D_{1ha}) that we need for our formula above, we know that SE($\hat{D}_{1ha}$)=1.1.

So,
$$\text{var}(\hat{D}_{1ha}) = \left[SE(\hat{D}_{1ha})\right]^2 = 1.1^2 = 1.21$$

Now, we place the value of 1.21 in our formula, and we find

$$\hat{\text{var}}(\hat{D}_{10ha}) = 100\,\hat{\text{var}}(\hat{D}_{1ha}) = 100 \cdot 1.21 = 121$$

Thus,
$$SE(\hat{D}_{10ha}) = \sqrt{\text{var}(\hat{D}_{10ha})} = \sqrt{121} = 11$$

In this case—after going through all of this work—the variance of the density is obtained in similar fashion to the density estimate. We multiply by 10 for both. But, it's good to check that intuitive result by doing the math!

For a last example, let us solve the problem we noted at the beginning of the chapter— extrapolating a daily survival estimate for bird nests to a 24-day estimate of the probability that a nest is successful and fledges at least one juvenile. Our daily survival rate estimate ($\hat{S}_D$) is 0.985, and our SE of the estimate is 0.023.

We will follow the 4-step process, again:

Step 1: write down the equation (function) of what you want to estimate:

Figure 5.4: *An ecologist might wish to calculate the probability of a wood thrush nest (left) surviving 24 days to fledge juveniles (right). The daily survival probability for the nest can be used to estimate 24-day nest survival. Photos by L. Powell.*

$$\hat{S}_{24-day} = \hat{S}_D^{\,24}$$

$$\hat{S}_{24-day} = 0.985^{24} = 0.696$$

Step 2: write out variance of new parameter in "delta method equation form":

$$\hat{\text{var}}(\hat{S}_{24-day}) = \hat{\text{var}}(\hat{S}_D)\left[\frac{\partial \hat{S}_{24-day}}{\partial \hat{S}_D}\right]^2$$

Step 3: find the derivatives you need for your function (*Note: this is in the form x^c in our table*):

$$(S_D^{24})' = 24 S_D^{23}$$

Step 4: arrive at equation you need:

$$\hat{var}(\hat{S}_{24-day}) = \hat{var}(\hat{S}_D)(24\hat{S}_D^{23})^2$$

which simplifies to:

$$\hat{var}(\hat{S}_{24-day}) = 576\,\hat{var}(\hat{S}_D)\hat{S}_D^{46}$$

Thus,

$$\hat{var}(\hat{S}_{24-day}) = 576 \cdot (0.023^2) 0.985^{46} = 0.1520$$

$$\hat{SE}(\hat{S}_{24-day}) = \sqrt{0.1520} = 0.390$$

We pause here and notice that the variance of our new parameter ($S_{24\text{-}day}$) is not intuitive with regards to the simple manner of extrapolating the daily estimate to the 24th power. Variance, and the SE, actually increase in magnitude because your uncertainty on a daily basis is compounded over the 24 days. Thus, the delta method was useful, as predicted.

Want more examples?

We direct you to an easy-to-use, on-line delta method calculator that performs more of these time-based transformations (e.g., monthly to weekly, annual to monthly, etc.). Powell (2007) has a table with ready-to-use formulas (see note on erratum in the References section, below). In addition, Powell (2007) works through other applications of the delta method in similar fashion to our coverage above.

Conclusion

It is common for ecologists to take their estimate of a parameter (e.g., survival, density, transition probability) and modify it in some manner—perhaps extrapolating across several time periods to change the scale of a survival estimate. Or, the parameter may be combined with another variable to calculate a new function. The delta method is a method that may be useful to approximate variance of the new function in situations similar to these. The user can follow 4 simple steps to obtain the formula for the variance approximation.

References

Cooch, E., and G. White. 2014. Appendix B: The 'Delta method' *In* Program MARK: a gentle introduction, 12th edition, Cooch, E. and G. White, eds. Online: http://www.phidot.org/software/mark/docs/book/pdf/app_2.pdf

Franklin, A. B., D. R. Anderson, R. J. Gutierrez, and K. P. Burnham. 2000. Climate, habitat quality, and fitness in Northern Spotted Owl populations in northwestern California. Ecological Monographs 70:539-590.

Hillborn, R., AND M. Mangel. 1997. The ecological detective: confronting models with data. Princeton University Press, Princeton, NJ. Larson, R. E., and R. P. Hostetler.

Mackenzie, D. I., J. D. Nichols, J. A. Royale, K. H. Pollock, L. L. Bailey, and J. E. Hines. 2006. Occupancy estimation and modeling: inferring patterns and dynamics of species occurrence. Elsevier Academic Press, Burlington, MA.

Powell, L. A. 2007. Approximating variance of demographic parameters using the delta method: a reference for avian biologists. The Condor 109: 949-954 (erratum at: The Condor 114(3):678). on-line delta method calculator from the author, available at: http://snr.unl.edu/powell/research/deltamethod_calculator.xls

Seber, G. A. F. 1982. The estimation of animal abundance and related parameters. 2nd ed. Chapman, London and Macmillan, New York.

Skalski, J. R., K. E. Ryding, and J. J. Millspaugh. 2005. Wildlife demography: analysis of sex, age, and count data. Elsevier Academic Press, Burlington, MA.

Ver Hoef, Jay M. 2012. Who Invented the Delta Method? The American Statistician, 66, 124-127.

White, G. C, D. R. Anderson, K. P. Burnham, and D. L. Otis. 1982. Capture-recapture and removal methods for sampling closed populations. Los Alamos National Laboratory LA-8787-NERP, Los Alamos, NM.

Williams, B. K., J. D. Nichols, and M. J. Conroy. 2002. Analysis and management of animal populations. Academic Press, San Diego.

Citing this primer

Powell, L. A., and G. A. Gale. 2015. Estimation of Parameters for Animal Populations: a primer for the rest of us. Caught Napping Publications: Lincoln, NE.

Ch. 6

Linear models[1]

"The shortest distance between two points is a straight line."
-- Archimedes

Questions to ponder:
- *How is a linear model used to describe variation in the value of a parameter?*
- *How do you put a discrete covariate in a linear model?*
- *What is a link function?*
- *How can we calculate the probability of survival from a set of estimated coefficients: B_0 and B_1?*

Thinking about parameters in linear models

You may be able to visualize the estimation of two survival rates for individuals in a sample—males and females, for example. It seems clear that we could divide the data into two groups—one for males and another for females. We could then estimate the survival rate for each of these sub-samples of the data.

In the language of parameter estimation, we would label "gender" in this example as a "covariate" of survival—we hypothesize that survival varies as gender varies. And, we note that "gender" is a discrete covariate—individuals are either male or female. We can place individuals into categories, based on the "gender" covariate.

In Chapter 4, we noted that modern software platforms for the estimation of parameters (e.g., MARK, SAS procs, R modules) provide the capability to assess continuous covariates such as mass, vegetation height, age (in days), body length, etc.. As such, we need to explore how this process works—as it is less intuitive than estimating survival for discrete groups, like males and females. *How can you estimate the probability of survival for individuals of a certain mass?*

To implement such analyses, we need to learn to view parameter estimation from the context of a **linear model**.

---

[1] *With thanks for content to Evan Cooch and Gary White*

Starting with the basics

If we return to what we know of the simplest linear model—the equation of a straight line—we know that models are often described, mathematically, as:

$$y = b + mx$$

This line describes how the value of y (our dependent variable) varies as the value of x changes. We may remember that a simple regression analysis would provide us with estimates of b, the intercept (where the line crosses the y-axis), and m, the slope (describes how steeply y changes with a change in x).

We need to modify the symbology used in this equation to match the manner in which quantitative biologists describe their models. So, we can write the same model using different symbols as:

$$Y = B_0 + B_1 x$$

Where

Y is the response variable, such as the probability of capture, occupancy, or survival
B_0 = the intercept
B_1 = the slope, or the coefficient for x

We can then imagine a scenario for which the estimated value of the parameter increases as mass increases—it might look like Figure 6.1. We know that the slope, B_1, is positive, and it seems that mass is an important covariate for the parameter. This model has one covariate (mass) and we will estimate two parameters for this analysis (the intercept, B_0, and the covariate for mass, B_1). Hence, $K=2$.

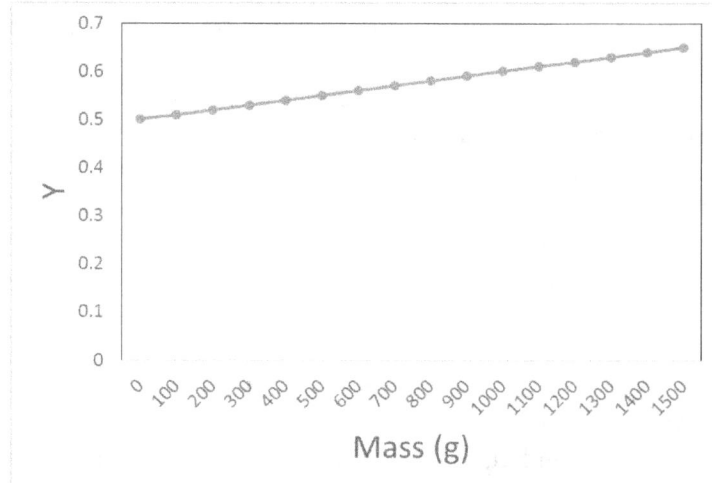

Figure 6.1: *Graphical illustration of the predictions of a hypothetical linear model showing a positive relationship between body mass and a response variable, Y.*

But, let's start even more simply: what does the linear model look like if mass does not affect the parameter value? The simplest linear model has only the intercept:

$$Y = B_0$$

And, if we plot this type of a model, we would see something like Figure 6.2 (at right)—showing that the value of the parameter is the same, regardless of mass. This model has no covariates, and there is one parameter estimated (the intercept): $K = 1$.

As you explore linear models, we suggest visualizing your analyses as figures to conceptualize the models you are using and creating.

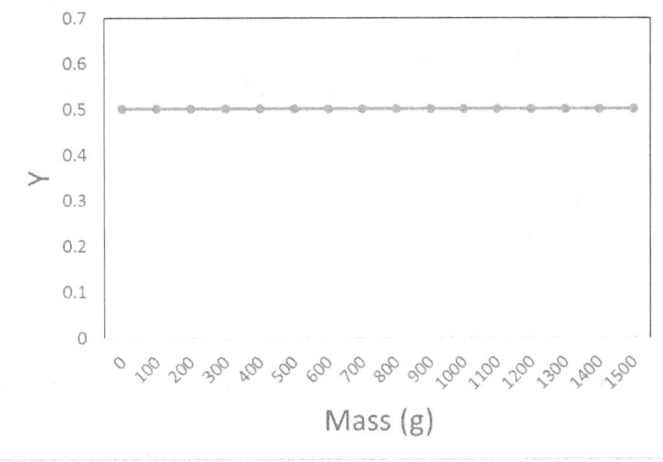

Figure 6.2: *Model predictions from a linear model with only an intercept ($Y=B_0$). Here, the response variable, Y, does not change with body mass. Such a model is often used as a 'null model' or 'intercept model' for comparison to more complicated models.*

Let's take a more complicated linear model—this time with two covariates: mass and age. Both covariates are continuous, ranging from 0 to some potentially infinite value.

The equation for this model would be:

$$Y = B_0 + B_1(mass) + B_2(age)$$

We will be estimating three parameters, the intercept and the two coefficients for our covariates. If we were to plug in values for mass and age, we might expect to be able to provide the following summary of our predictions for the parameter value.

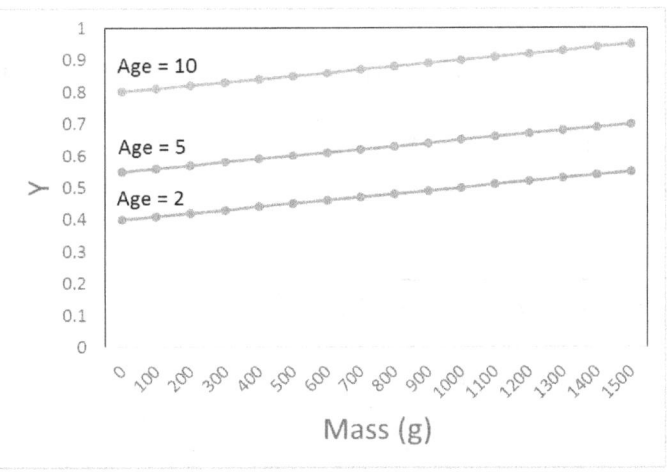

Now, we can see that both mass and age of the individual are important. Our estimate of the parameter value, Y, is going to vary depending on the mass and age. You should be able to see that for individuals with a mass of 800 g, the oldest individuals would be predicted to have higher survival than younger individuals.

Figure 6.3: *Model predictions for the value of a response variable, Y, as body mass increases. Predictions are shown for 3 different ages (age 2, 5, and 10).*

Dummy variables: for discrete variables

Now that we have conquered understanding of continuous covariates in linear models for parameter estimation, we need to come back and think about our simpler discrete models. How do we put a discrete covariate into a linear model?

Let's use the example of an effect of gender, as before. If we follow the general format of our continuous models, the model would look something like:

$$Y = B_0 + B_1(\text{gender})$$

And, that is correct! However…how do we put "male" and "female" into the model to calculate a value for the parameter, Y? The answer is—we create a "dummy variable".

"Dummy variables" are used to code discrete, or categorical, covariates. Once a covariate is selected, such as gender, we establish a binary code for the states of the covariate. For example, **we could use "0" to code for male and "1" to code for female.**

Thus, if we insert the value of the code for gender in our linear model, we get the following models:

Male: $Y = B_0 + B_1(0)$
Female: $Y = B_0 + B_1(1)$

You will note that because the "0" we insert for males eliminates the value of the covariate for gender ($0*B_1 = 0$), we end up with the intercept of the model (B_0) as the estimate for the parameter. Further, the value of the parameter for females is the value for males, plus or minus an amount controlled by the coefficient for gender. In essence, we are estimating a "female effect." Is the parameter estimate higher for females (B_1 for females >0) or lower (B_1 for females <0)?

We've used one dummy variable and the "0" and "1" coding to create a linear model for a discrete covariate with two states, such as male and female for gender. But, how do you treat a categorical covariate such as month—perhaps we have three values in our study: May, June, and July.

Use dummies! Don't be a dummy!

We stress that some discrete covariates "appear" to have numerical values, and some readers may be tempted to use the values as a continuous covariate. For example, **months** of the year can be given values from 1-12. However, **under no circumstances** should you create a model with a continuous covariate using the numbers associated with each month, such as:

$$Y = B_0 + B_1(\text{month})$$

Months are, of course, discrete units. June does not really equal "6", nor does July really equal "7". Further, July is not 1 unit larger than June—and December (month #12) is not twice as large in magnitude as June (month #6). Most importantly, the probability of survival should not be expected to increase in proportion to an increase of 1 'month unit'. Rather, **we should use dummy variables to create an appropriate linear model.**

In our previous example with gender, we had two states of the discrete covariate—and thus, we were able to use "1" and "0" to code for both states with a single covariate. Now, we have three states—the three months. We solve the problem by using two dummy variables as such:

State of discrete variable: *Month*	Covariates used	
	June	July
"May"	0	0
"June"	1	0
"July"	0	1

We note that it appears to be a **general rule for dummy variables** (and, yes—it's true!):

If we have n states of a discrete covariate, we must have n-1 dummy variables to code for the states.

We suggest using values of 0's for the first state—which serves as a baseline for the other states. Here, we establish two covariates "June" and "July". When we code the June column, we ask the question, "Is this state June"? If not, we place a "0". If yes, we place a "1". Similarly for the July column, we ask, "Is this state July?" If so, "1". If not, "0". "May", as indicated by the two 0's in the two columns for June and July is neither "June" nor "July".

The linear model, then, has 3 parameters—an intercept and the two covariates (June and July):

$$Y = B_0 + B_1(\text{June}) + B_2(\text{July})$$

To predict the value of the parameter for May, we insert the 0's as such:

$$Y_{May} = B_0 + B_1(0) + B_2(0)$$

Again, we note that the value of the parameter is the value of the intercept, as the June and July coefficients become 0—leaving us only with the intercept.

Similarly, for June and July:

$$Y_{June} = B_0 + B_1(1) + B_2(0)$$
$$Y_{July} = B_0 + B_1(0) + B_2(1)$$

And, we can again see that we are essentially estimating a "June" effect and a "July" effect for our parameter. How much higher or lower is the estimate, relative to our **baseline estimate** (the intercept) for May?

You try it

Suppose you are interested in estimating survival as a function of the land cover in which animals were living. You have 5 states for land cover: wetland, prairie, forest, crop fields, and urban. How many dummy variables would you use? Can you create a table similar to the table above for this problem? Can you write out the linear model that predicts survival in forest?

Link functions

We should now be comfortable thinking about parameter estimation in the context of a linear model. As an example, let us return to the interest in describing variation in survival by the mass of the animal and the age of the animal (both continuous covariates):

$$S = B_0 + B_1(mass) + B_2(age)$$

As we prepare to estimate the coefficients, B_0, B_1 and B_2, we find that we are not performing a standard regression analysis. The first clue that we will need some special treatment is that we can see that our response variable, **S, or survival, needs to be bounded between 0 and 1**—that is, survival is a probability.

To ensure that $0 \leq S \leq 1$, we will need to use a link function. A link function literally links the response variable (alive or dead) with the explanatory variable, such as mass or age in our example. We will estimate our parameter as a transformed variable, and then we will back-transform to obtain our survival estimate.

Some **common link functions** used for parameter estimation include the following:

sin
logistic (or logit)
log

Note that the sin and logit link functions act to constrain the parameter value within the space 0.0 to 1.0. The log link function does not. For that reason, you may find that software such as program MARK suggest the log function when estimating population size (which does not have to be constrained between 0-1).

You should notice that we have been using the form of the linear model, for our previous discussions:

$$Y = B_0 + B_1(mass) + B_2(age)$$

But, when using the logit link function, the linear model is actually:

$$\log\left(\frac{\gamma}{1-\gamma}\right) = \beta_0 + \beta_1(mass) + \beta_2(age)$$

Hence, the coefficients B_0, B_1, and B_2 are estimated as logit-scale coefficients. To make sense of them, and to use a linear model to predict the probability of survival under various values of the covariates, we must back-transform the logit-scale function to the "real world". We do this with the equation:

$$\gamma = \frac{\exp^{(\beta_0 + \beta_1(mass) + \beta_2(age))}}{1 + \exp^{(\beta_0 + \beta_1(mass) + \beta_2(age))}}$$

Dipper example

Let's use a simple example, provided by Cooch and White (2014)—the infamous European dipper (*Cinclus cinclus*) data set used by Lebreton et al. (1992). Cooch and White (2014) used this example to show beginning users of program MARK how to perform an analysis of CJS-type data (Chapter 10), and the software provides an estimate of annual survival under a null, constant-survival model, $S = 0.5658$.

But, how did the software obtain that estimate? Program MARK has estimated the survival estimate through the use of a simple linear model with a "sin" link function and no covariates:

Figure 6.4: *European dipper (Cinclus cinclus), Kirkcudbright, Scotland. Photo by Mark Medcalf, and available in the public domain.*

$$sin\ S = B_0$$

Our software provides the maximum likelihood estimate of B_0, using the sin link function:

$$B_0 = 0.13203.$$

Because the coefficient, B_0, is a sin-scale coefficient, so we must back-transform it to make sense of it as a survival probability.

We do this with the general equation, which transforms sin-scale linear models to real-life survival probabilities that are constrained between 0.0 and 1.0:

$$S = \frac{\sin(B_0) + 1}{2}$$

Hence:

$$S = \frac{\sin(0.13203) + 1}{2} = 0.56582$$

That survival probability of $S=0.56582$ matches the earlier estimate that the software provided. And, we can now see how we can work back from a sin-scale coefficient to predict the probability of survival. This will be important when we begin our analyses and find coefficients (e.g., B_1) that are on a logit- or sin-scale.

Ring-necked pheasant nest survival example

The previous dipper example allowed us to practice back-transforming a linear model from the sin scale to a survival probability. But, it was the simplest type of linear model that only had an intercept and no covariate. The survival rate, then, was constant across space and time. What if survival varies?

To illustrate how to work with a more complex linear model, we will add one covariate. The ring-necked pheasant (*Phasianus colchicus*) is a species of economic importance in the central United States, although it is an introduced species. Landowners have interest in managing the species for hunting opportunities.

Matthews et al (2012) conducted a nest survival study, so the data were considered "known fate" (Chapter 9). In some initial investigations, the authors considered a model to evaluate the effects of bare ground on the daily probability of nest survival. Bare ground is measured as the "percent cover" of bare ground in a quadrat surrounding the nest site, and the average value for percent bare ground in their study was 8%.

The authors used the logit link function for their analysis, so the model was:

$$\log\left(\frac{S}{1-S}\right) = \beta_0 + \beta_1(\%\_bareground)$$

The estimates for the coefficients, B_0 and B_1, were:

B_0 = 3.4200653 (SE = 0.27)
B_1 = 0.0336702 (SE = 0.03)

How do we obtain estimates of survival along a gradient of bare ground conditions? For this example, we will ignore the standard error so that we can focus on our objective to show how to estimate nest survival with a single covariate. An astute reader may see that the standard error for the bare ground coefficient (B_1) is very close to the value of the coefficient (both are equal to approximately 0.03), which suggests that this effect may not be the most important factor to explain variation in survival of pheasant nests—but stay with us as we follow this as a simple example of a covariate in a linear model.

The first step to estimating survival at various levels of bare ground is to create a table, and we suggest using a spreadsheet to help with the calculations—although you can certainly do them by hand or with a calculator if needed. Our table will have three important columns. The first is the values of bare ground that we wish to consider (we will use 0% to 12%). The second, is our intercept, B_0. And, the third is the coefficient for our variable of interest (bare ground, in our example), B_1. We can then fill those columns with the values that will be used in our equations. Our coefficients, B_0 and B_1 will be constant in all rows of the table, and the values for bare ground that we wish to consider are selected along a suitable range for the data.

% bare ground	B_0	B_1	logit(S)	S
0	3.420065	0.03367	3.420065	0.968326
2	3.420065	0.03367	3.487406	0.970327
4	3.420065	0.03367	3.554746	0.972206
6	3.420065	0.03367	3.622087	0.973969
8	3.420065	0.03367	3.689427	0.975623
10	3.420065	0.03367	3.756767	0.977174
12	3.420065	0.03367	3.824108	0.978629

Now that we have our first three columns populated with numbers, we can start our calculations for S. We have highlighted four cells in the table above, logit(S) and S for the bare ground level of 0%, and logit(S) and S for the bare ground level of 10%. How are those values calculated?

We start with the logit(S) column, which is the logit-scale estimate of survival, and we remember our equation:

$$\log\left(\frac{S}{1-S}\right) = \beta_0 + \beta_1(\%\_bareground)$$

Therefore, for the level of bare ground of 0%, we calculate:

$$\log\left(\frac{S}{1-S}\right) = 3.420065 + (0.03367 * 0) = 3.420065$$

And, for the level of 10%, we calculate:

$$\log\left(\frac{S}{1-S}\right) = 3.420065 + (0.03367 * 10) = 3.756767$$

In similar fashion, we can fill in all other values of logit-scale survival for our range of values for bare ground. But, obviously, survival probability must be constrained between 0 and 1, so our next step is to back-transform our logit-scale survival to a probability by adapting the formula we used earlier in this chapter:

$$S = \frac{\exp^{(\beta_0 + \beta_1(\%\_bareground))}}{1 + \exp^{(\beta_0 + \beta_1(\%\_bareground))}}$$

For our values of 0 and 10, we would calculate daily survival as:

$$S = \frac{\exp^{(3.420065)}}{1 + \exp^{(3.420065)}} = 0.968326 \qquad S = \frac{\exp^{(3.756767)}}{1 + \exp^{(3.756767)}} = 0.977174$$

And, we can finish the table with values for all other levels of bare ground. At this point, we can see the daily nest survival does increase, gradually, as bare ground increases (Figure 6.5). But, survival only increases by approximately 1% as bare ground increases from 0% to 12%, which confirms our initial assessment from inspection of our model coefficients and standard errors.

You may also detect a faint curve in the line. You might ask, "How that is possible if this is a linear model?" And, it is true that we used a linear model to estimate our coefficients—but we did that in the logit-scale. When we back-transform, our linear model often appears non-linear with a slight curve because of the transformation.

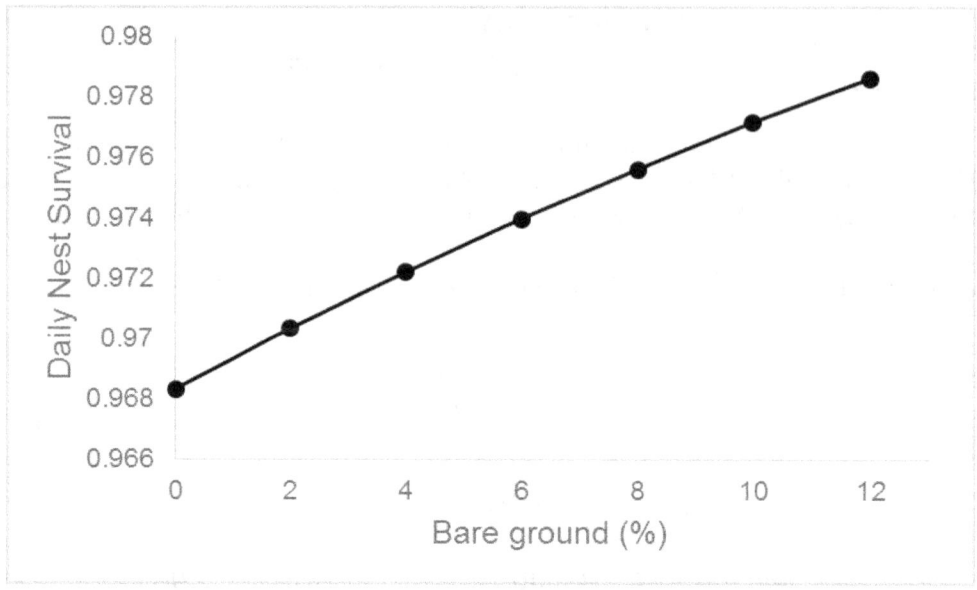

Figure 6.5: *Change in estimate of daily nest survival for ring-necked pheasant across a gradient of cover (%) of bare ground.*

Conclusion

The ability to conceptualize parameter estimation in the context of linear models—and the ability to write out the linear models for our analyses—takes our skill-set to a new level. To use linear models to estimate parameters such as probability of survival or capture, link functions are needed to constrain the resulting estimates between 0 and 1. We must use a back-transformation to predict the probability of survival rate after we are provided estimates of logit- or sin-scale coefficients for our linear models.

References

Cooch, E., and G. White. 2014. Chapter 6: Adding constraints: MARK and linear models. *In* Program MARK: a gentle introduction, 12th edition, Cooch, E. and G. White, eds. Online: http://www.phidot.org/software/mark/docs/book/pdf/chap6.pdf

Matthews, T. W., J. S. Taylor, J. S., and L. A. Powell. 2012. Mid-contract management of Conservation Reserve Program grasslands provides benefits for ring-necked pheasant nest and brood survival. Journal of Wildlife Management 76:1643-1652.

Lebreton, J. D., K. P. Burnham, J. Clobert, and D. R. Anderson. 1992. Modeling survival and testing biological hypotheses using marked animals: a unified approach with case studies. Ecological Monographs 62: 67-118.

For more information on topics in this chapter

Conroy, M. J., and J. P. Carroll. 2009. Quantitative Conservation of Vertebrates. Wiley-Blackwell: Sussex, UK.

Donovan, T. M., and J. Hines. 2007. Exercise 4: Single-species, single-season model with site level covariates. *In* Donovan, T. M., and J. Hines. Exercises in occupancy modeling and estimation. On-line: http://www.uvm.edu/envnr/vtcfwru/spreadsheets/occupancy/occupancy.htm

Williams, B. K., J. D. Nichols, and M. J. Conroy. 2002. Analysis and management of animal populations. Academic Press, San Diego.

Answers: You try it!

State of discrete variable: *land cover*	covariates used				
	prairie	forest	crop	fields	urban
wetland	0	0	0	0	0
prairie	1	0	0	0	0
forest	0	1	0	0	0
crop	0	0	1	0	0
fields	0	0	0	1	0
urban	0	0	0	0	1

$Y_{Forest} = B_0 + B_1(0) + B_2(1) + B_3(0) + B_4(0) + B_5(0)$

An aluminum leg band is used to tag a cliff swallow (Petrochelidon pyrrhonota) in a large mark-recapture study in Nebraska, USA. Biologists have conducted many analyses that incorporate linear models to detect sources of variation in probabilities of movement and survival for this species. Photo provided by Mary Bomberger Brown and used with permission.

Ch. 7

Introduction to mark-recapture models

"Man is the only kind of varmint who sets his own trap, baits it, and then steps in it."
-- John Steinbeck

Questions to ponder:
- *What is the difference between closed and open populations?*
- *What is meant by 'encounter probability'?*
- *How do catch-per-unit-effort methods compare to mark-recapture methods?*

Mark-recapture data and analyses

The use of mark-recapture data has a long history in wildlife management and ecological studies. The placement of marks on animals will continue to be a staple method in the toolkit of a wildlife biologist. Mark-recapture data allows us to answer questions such as: "how many animals are there in this population?" Or, "Are some animals at a higher risk of mortality than other animals?" Or, "what is the probability that animals in the population will remain in the population?"

As technology has changed, biologists have added more type of marks to their toolkit—for example, the use of Passive Integrated Transponders (PIT) and satellite telemetry. But, the basic approach to analysis of mark-recapture data has not changed in decades. Here, we will explore some newer methodologies and analysis models, but we begin with the standard: a "closed" mark-recapture model (Chapter 8) to consider animal abundance questions. Chapters 9-13 will explore "open" mark-recapture models to address questions of mortality and movement.

The common denominator of all mark-recapture methodology is that an animal is captured in some manner and marked. The animal is then released and captured at a subsequent time period. Some marking methods provide unique identifications (e.g. "pheasant #133"), but other marking methods only show than an animal is marked (e.g., a splotch of white paint on the scales of an iguana).

Samples and encounter probability

We know that when we capture animals in most populations, we are not able to capture the entire population at one time. Thus, by definition, we are sampling the population. And, we will try to make inferences about the population from our sample.

What if you were told that a biologist went to three study sites, placed some sort of net or trap, and recorded the capture of 100 unique individuals at each site? Could you make an inference about the size of the population without knowing any other information? The inference, certainly, is limited—the best inference is that there are at least 100 individuals in each population.

But, how many animals are really in the population? What additional information about our trapping effort do we need to know?

We must know something about the capture, or encounter, probability. That is, what is the probability that an individual living in the population is captured? In mathematical language, we have our sample, n, but we want to estimate the population size, N. We know that if capture probability is p:

$$\hat{N} = \frac{n}{\hat{p}}$$

So, if we told you that in the first population, $p = 0.30$, you could tell me that the population size is 333. We know that because we captured 30 percent of the population (another way of saying that an animal had a 30% chance of being captured in our trapping program). So, 100 animals represented 30 percent of the population. And, by math we can see that $\hat{N} = 100/0.30 = 333$.

> ### Types of encounters
>
> Specific types of research often have different terms to indicate the type of encounter that is involved:
>
> **"Capture probability"** typically refers to the probability that an individual is captured during a given time period in an actual trap or net.
>
> **"Recovery probability"** is defined as the probability that a hunter will, during a given time period, shoot a marked animal—shot and recovered animals cannot be released to the wild because they are dead. Thus, "recoveries" are a special type of mark/recapture data.
>
> **"Re-sighting probabilities"** are defined as the probability that an animal marked with colored or numbered tags will be seen by a person and recorded (i.e., not in a trap) during a given time period.

Can you estimate the population size if $\hat{p} = 0.50$? What if $\hat{p} = 0.70$?

You should find that $\hat{N} = 100/0.50 = 200$ and $\hat{N} = 100/0.70 = 142$.

So, capturing 100 animals in each population is only our sample, and we cannot infer much about the population size until we know the encounter probability, p.

NOTE: You may see, especially in older literature, the term 'encounter rate' or 'capture rate'. Most scientists (and editors of journals!) now prefer the term "encounter probability" as a true "rate" has a time-specific element attached, such as a flow rate of a stream (meters/second). Although it is true that 'survival rates' also have a time element that should be specified (annual, monthly, daily, etc.), the stochastic nature of binomial-type events is emphasized with the use of the term "probability".

Catch-per-unit-effort

There are implications to standard methodologies that assume equal catchabilities—fisheries biologists and small mammal biologists, for example, historically calculated "Catch-per-unit-effort" (CPUE) statistics. CPUE might be expressed as 350 fish/net or 1300 small mammals per trap night.

But, if a net captures 350 fish of size >=700mm, 350 fish of size between 200 and 500mm, and 350 fish of size <200mm, we are in the same situation as we were with our 100 animals captured at each of three study sites. The size of the holes in the net, perhaps, might favor the capture of fish in the 200-500 mm size—larger fish bounce off and small fish swim through it. So, if capture probability is highest for the 200-500mm fish and lowest for the other two size classes, our CPUE cannot be used to infer anything about population size. In fact, there would be many more of the smallest- and largest-sized fish, in our example, than the middle size class. Why? Because capture probability of the smallest and largest size classes was very low. For example, if $p=0.10$ for those size classes, $\hat{N} = 350/0.10 = 3500$ fish. And, if $p=0.9$ for the middle size class, $\hat{N} = 350/0.9 = 389$ fish.

Despite the original inference from CPUE of similar numbers of each size category, we now see that there are many more small and large fish than mid-sized fish. For this reason, most fisheries biologists use CPUE results with caution because they know that their nets' capture efficiencies are different for individuals of different sizes.

Some logic

Before we launch into the mathematics of the simple estimator for population size, let's take a look at the logic behind the estimator. In Figure 7.1, you see two populations (one at top, the other at the bottom). A sample of 100 animals is taken during the first capture occasion in both populations. Things are looking fairly even at the moment. But, we should be aware that we can't infer much about population size, yet. We only know that each population has at least 100 animals!

Next, a second capture occasion is held and another 100 animals are captured (perhaps our traps or nets only hold 100 animals?). But, some interesting differences arise, now. In the first population only 20 of the 100 animals captured in time 2 are marked. And, in the second population, 80 of the 100 animals captured in time 2 are marked (Figure 7.1).

Can you suggest which population is larger than the other?

You should have answered "Population #1" (Figure 7.1). Why? Because we marked 100 animals in both populations in time 1. And, in time 2, we appear to have marked almost all (80% in fact) of Population #2, while we appear to have only marked 20% of Population #1. If 100 animals are only 20% of the population in Population #1, it would seem to be very big—much bigger than Population #2. In a few pages, we will put some equations to this logic. For now, you should be able to look at this example and see we can make a much better inference about population size with a very simple mark-recapture experiment.

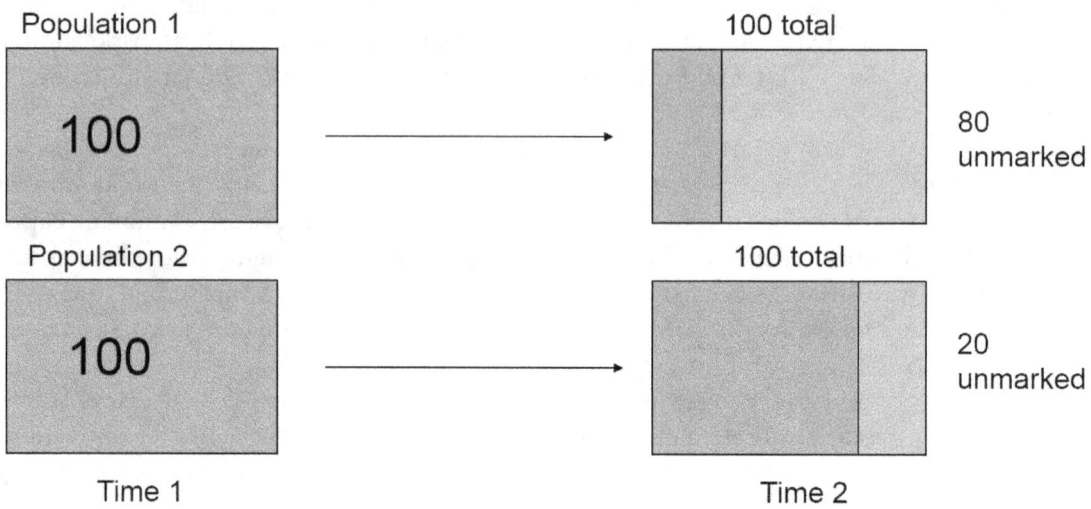

Figure 7.1: *Two populations with very different sizes (Population 1 and 2) are sampled with 100 animals marked in each population during Time 1. During Time 2 another 100 animals are sampled from each population. The number of marked animals captured varies between populations in Time 2. The presence of many unmarked individuals in Population 1, relative to the number of unmarked individuals in Population 2 suggests the Population 1 is bigger than Population 2.*

The cohort

An important concept in the use of mark-recapture analyses is the "cohort". A cohort is the number of marked animals released during time period t. We use cohorts to establish capture histories, which are the basis for all analyses of mark-recapture data. Capture histories are records that show when an individual has been captured during a study. For example, if a cohort of individuals is captured during time period 1 and then released immediately, those animals can have four different capture histories during a study with three capture occasions: "111", "110", "101", and "100". For simple capture histories, we use a "1" to indicate that an animal has been captured (a 'success'). And, we use a "0" to indicate that an animal was not captured (a 'failure').

This type of data is called **"binomial"**—two forms (1/0, or success/failure).

But, the released cohort is critical to our analyses—all subsequent work is conditioned (or statistically based) on that initial sample of animals that are captured and released. Mark-recapture analyses, at their core, are an attempt to estimate something you don't know (e.g., population size). To accomplish that goal, you must have something you do know—and when we release a cohort of animals, we know how many we released. And, that will be very important, as you will see.

For example, if you were to decide to study ducks in the state of New York, USA, you could not start by collecting bands, or rings, or ducks recovered by hunters in New York. Those bands, or rings, would provide useless data unless you knew how many birds had been banded, or ringed, initially. That is, without knowledge of the cohort, all you have is a shiny pile of bands in your hand!

Closed or open?

Mark-recapture analyses can be divided into two main types of analyses: analyses of 'closed' populations and analyses of 'open' populations.

Remember the basic equation of population growth (Chapter 2) is a function of births (B), deaths (D), immigration (I), and emigration (E): $\Delta N = B + I - D - E$.

We refer to 'closed' populations as having $\Delta N = 0$. That is, no births, deaths, or movements of animals in or out. Although we know that no species exists in closed populations, populations may be considered 'closed' if we study them over very short time periods. This is necessary to estimate population size. And, it is important to note that 'very short time period' is relative to each species. The scale over which animals are born and die is short for small rodents, relative to elephants. So, a month-long study of small rodents might be considered to be too long for our closure assumption (many small rodents might die or be born during a month). But, a 4-month study of elephants might be considered very acceptable (very few deaths and births are likely during such a period).

We will come back to questions of closure of populations later, but this is a key consideration when you start to design a study for your species. When are the birth pulses? How often do individuals die? After a given time period (e.g., day, week, month), will animals have left your study site?

Alternatively, models of "open" populations are often more realistic. When we study 'open' populations, we assume that deaths are occurring. Mark-recapture methods treat movements in unique ways, but many types of analyses of open populations assume movements are occurring as well.

Therefore, studies of open populations are usually longer in duration than studies of closed populations (however, remember that a closed study of elephants might be longer than an open study of mice!). In fact, we want deaths to occur during studies of open populations. If animals do not die (or, for more complex models, leave or enter the study site), our study will not be able to estimate a useful survival (or movement) probability.

Models matter

As we start to envision our analyses as models of systems, it may be useful to compare simple open and closed model structures. As you begin working with mark-recapture data, you must start to think in terms of probability statements. So, let's start with a simple example:

Let's consider the situation where you captured a sample of individuals and released them as a cohort during time 1. Then, during time 2 you recapture some of that cohort. Given that you captured the animals during time 1, they will either have a "11" or "10" capture history (obviously, other animals not captured during time 1 could have a "01" or a "00" capture history as well, correct?).

So, what is the probability of having a capture history of "11"? It depends completely on the 'model' under which we are working. Is it a closed system? If so, the only thing that had to happen for our animal released in time 1 to have a "11" capture history is…to be captured again. So, we can represent that as a probability statement: p_2—the probability of capture during time 2.

What is the probability of having a 10 capture probability? It is $1-p_2$, or the probability of NOT being captured during time 2. It is as simple as that.

Now, let's change our model. Let's say, instead, that we are doing an analysis of animals in an open population (specifically, a "Cormack-Jolly-Seber," or CJS-type, open population—more on that in Chapter 8). We have the

Type of model	Capture history	Probability of animal having capture history
Closed population	11 10	p_2 $1-p_2$
Open population, CJS	11 10	$S_1 p_2$ $S_1(1-p_2)+(1-S_1)$

Note: probabilities for capture histories are conditioned on capture and release in time 1

same animals released in time period 1. Now…what has to happen for them to have a capture history of "11"? It's going to be more complicated than the probability in our closed model—because there is something else happening: *some animals are dying.*

For an animal to have a capture history of "11" in an open population, the animal must first survive the interval in which they are released, and then they must be captured in the next interval. So, the probability would be: $S_1 p_2$. We multiply the two probabilities, S_1 and p_2, together, as the animal must do both—survive and be captured. We will cover this again when we discuss open populations, but we assume in analyses of open populations that the cohort of animals is released simultaneously at the beginning of time 1. So, they must survive through time period 1, and hence we used the S_1 designation to refer to survival during time period 1. Capture happens at the beginning of the next time period, so we use the p_2 to designate capture probability in time 2.

What is the probability of having a capture history of "10"? You might simply look at the closed model and suggest: $S_1(1-p_2)$. And, you would be partially correct. That is, an animal *can* have a capture history of "10" by surviving time 1 and then avoiding capture in time 2.

But, there is *another alternative* that also results in a capture history of "10"—the animal might die during time 1. Yes, as the famous Alfred, Lord Tennyson wrote, Nature is *"red in tooth and claw."* And, that affects our statement of capture probability—we must account for both possibilities. The total probability for having capture history of "10" is the *sum* of the two alternatives: capture avoidance and death. So, we write it as: $S_1(1-p_2) + (1-S_1)$. For a further discussion of CJS-type capture histories with more than two capture periods see Chapter 10.

Conclusion

Biologists use mark-recapture models to estimate parameters for populations, and the models are based on probability statements that require certain assumptions about the population in question. To estimate population size, we use models that assume a 'closed population' (no deaths or movement in/out). To estimate survival, we use models that are based on 'open populations'. Capture histories (e.g., "1001100") are the typical type of input data for these analyses, and our analyses are conditioned on the released cohort: animals released at the same time.

For more information on topics in this chapter

Amstrup, S. C., T. L. McDonald, and B. F. J. Manly. 2005. Handbook of capture-recapture analysis. Princeton Univ. Press: Princeton, NJ.

Conroy, M. J., and J. P. Carroll. 2009. Quantitative Conservation of Vertebrates. Wiley-Blackwell: Sussex, UK.

Cooch, E., and G. White. 2014. Chapter 1: First Steps. *In* Program MARK: a gentle introduction, 12th edition, Cooch, E. and G. White, eds. Online: http://www.phidot.org/software/mark/docs/book/pdf/chap1.pdf

Nichols, J. D. 1992. Capture-recapture models. BioScience 42: 94-102.

White, G. C., D. R. Anderson, K. P. Burnham, and D. L. Otis. 1982. Capture-recapture and removal methods for sampling closed populations. Los Alamos National Laboratory. LA-8787-NERP. 235 pp.

Williams, B. K., J. D. Nichols, and M. J. Conroy. 2002. Analysis and management of animal populations. Academic Press, San Diego.

A female white-tailed deer (Odocoileus virginianus) is marked with a numbered ear tag and a radio-telemetry collar in western Iowa, USA. Photo by Greg Clements, University of Nebraska-Lincoln.

Ch. 8

Estimation of population size: closed populations

"My parents removed to Missouri in the early 'thirties; I do not remember just when, for I was not born then and cared nothing for such things...[our] home was made in the wee village of Florida, in Monroe County, and I was born there in 1835. The village contained a hundred people and I increased the population by 1 per cent. It is more than many of the best men in history could have done for a town. It may not be modest in me to refer to this, but it is true. There is no record of a person doing as much-- not even Shakespeare. But I did it for Florida, and it shows that I could have done it for any place--even London, I suppose."
-- Mark Twain

Questions to ponder:
- *Is there a fixed period of time that should be considered 'closed' for all species?*
- *What kind of information is needed for a simple Lincoln-Petersen estimate of N?*
- *How can I assess the potential for bias with a simple Lincoln-Petersen estimate when I have small samples?*

A bit of history (*histoire*)

The Lincoln-Petersen estimator was first used to estimate the population size of a **closed** wildlife population in 1896 when **C. G. Johannes Petersen**, a marine biologist in Denmark, estimated the size of a population of plaice (*Pleuronectes platessa*)—a species of flatfish, similar to flounder (Petersen 1896).

Frederick C. Lincoln, an American ornithologist, described the method in 1930 in USDA Circular 118: "*Calculating Waterfowl Abundance on the Basis of Banding Returns.*" Lincoln, who also developed the 'flyway' concept for migratory birds, devised the method to estimate the continental population size of waterfowl (Lincoln 1930).

Although we give Lincoln and Petersen credit for this method, the general idea of using a 'known ratio' to estimate components of an 'unknown ratio' is much older. Ken Pollock, of North Carolina State University, suggests that the first use of the Lincoln-Petersen-type estimator was by a chap named **Laplace** in 1783. Laplace wanted to estimate something useful—the population of France. There were a lot of people in France, and census-type data were not common at this time. But, one thing was known—the number of births in the whole of France. So, Laplace knew how many babies (think of babies as 'marked animals' in time 1).

$$\hat{N} = \frac{n_1 n_2}{m_2}$$

In a small subset of the parishes (a small, local geographic region similar to 'counties' in the US) in France, Laplace could obtain a relatively accurate count for the total number of people in the parishes (think of this as n_2), and he was also able to obtain the number of babies born in those same parishes (again, babies are 'marked individuals', so think of this as m_2).

So, he was ready to apply the estimator. If he assumed that the birth rate (babies/population) was the same in those parishes as it was in the whole of France, he could estimate the population of France (N) as equal to $n_1 * n_2/m_2$ (Laplace 1783).

Perhaps it is useful for you to see that this estimator has been used for a very long time?!

Closed population models

In Chapter 7, we compared open and closed population models. Here, we will continue with the topic of closed populations. Thus, we will assume no births and no deaths during the sampling of our population. The simplest form of a closed mark-recapture analysis is called the **Lincoln-Petersen** method. Although it is simple, the Lincoln-Petersen method provides an unbiased maximum likelihood estimate of N for a two-occasion sample. The underlying assumption is that the proportion of marked animals remains the same during the two sample periods.

The L-P method begins with the basic idea that an unknown portion of the population is captured and marked during the first time period. And, although you don't know what that proportion is, we can denote it as the capture probability, p. That is, the unknown proportion of the population that is marked during the first time period is equal to the capture probability (if p = 0.35 during time 1, then you should capture 35 percent of the population).

But, we obviously don't know anything except how many animals we captured. This sounds like the example earlier in this chapter of 100 animals captured in a study area, doesn't it? Well, never fear, Lincoln and Petersen both came up with a brilliant plan to estimate p, and therefore estimate N.

The brilliantly simple idea is that if you have a second capture occasion and you look around at all of your captured animals, the proportion of the animals with marks in your sample SHOULD be the same proportion of the population that you captured (and marked) during time 1. So simple! We can see this graphically in Figure 8.1.

If those proportions are equal, then we can use an equation to state:

$$\frac{m_2}{n_2} = \frac{n_1}{N} \quad \Longrightarrow \quad \hat{N} = \frac{n_1 n_2}{m_2}$$

That is, the portion of animals caught (n_1) from the population (N) during the first time period is equal to the portion of the sample during time 2 (n_2) that have marked on them (m_2). By rearranging with algebra, we can solve for the only unknown in the equation, which is the parameter whose value we want to estimate: N.

Lincoln and Petersen's method is simple, and has a few **assumptions**. First, we assume that capture probability is the same for all animals in the population. Interestingly, p_1 and p_2 do not have to be the same, nor do the capture methods need to be the same. In fact, Lincoln used traps to band and mark ducks and then he used hunter kills as the second capture method.

Other assumptions:
- the marking of an animal does not affect p_2.
- marks are not lost between capture occasions.
- all marks at time 2 are reported.
- each sample of the population (during time 1 and 2) is a random sample of the population.

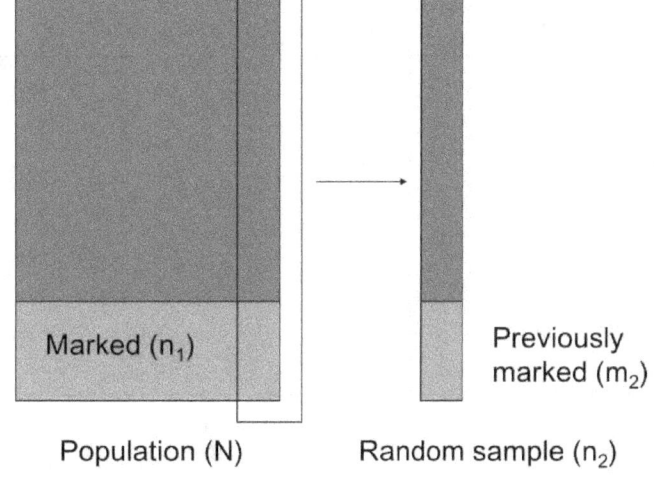

Figure 8.1: *The logic behind the 2-sample Lincoln-Petersen approach to estimation of population size. First (left), a population of unknown size (N) is sampled and n_1 individuals are marked (the unmarked portion of the population is represented by dark color). A second random sample (n_2) of the population is taken (right). The proportion of marked animals can be determined in the second sample, and it should be equal to the proportion of marked animals in the population immediately after the first sampling/marking.*

You try it!

Ready to try your hand at a problem?

Pollock et al. (1990) described a study in Florida in which 148 northern bobwhite (quail [*Colinus virginianus*]) were marked with leg bands after being trapped in cornfields during January of 1982. During a 3-day controlled hunt in February 1982, 39 of 82 shot quail were determined to have leg bands.

What is the estimate of the population size ($\hat{N}$) by the Lincoln-Petersen estimator? To provide the estimate, you need to figure out the values for n_1, n_2, and m_2. You'll find the answer at the end of this chapter.

Generalizing the Lincoln-Petersen

Thus far, we have used summaries of n_1, n_2, and m_2 to estimate population size. But, we need to be working towards an understanding of the generalization of the method—how does it work when you have a set of capture histories that will be the input for a more complicated analysis in a software package such as Program MARK?

Four possible capture histories	Probability of having capture history	Number of animals with this capture history
10	$p_1 q_2$	$n_1 - m_2$
01	$q_1 p_2$	$n_2 - m_2$
11	$p_1 p_2$	m_2
00	$q_1 q_2$	$N-r$ (NOTE: *can't observe*)

The table above shows the four possible capture histories of animals in a population under the 2-sample Lincoln-Petersen research design. Note that the first three will be the animals in your data set, and the 4th capture history ("00") is not observable—these are the animals that are never captured.

We can also define the capture probability as p and the probability of not being captured as q (which is also, by definition, *1-p*). You have seen n_1, n_2, and m_2 before. The last definition we need is for r, which is the total number of captured animals. Although r is helpful, we are really interested in N, the population size.

If we allow capture probability to be different in each time interval (thus, the use of p_1 and p_2 to indicate that capture probability may be different in time 1 and time 2), we can see that the probability of having a capture history of 11 is: $p_1 * p_2$. That is, the probability of being captured during time 1 is p_1 and the probability of being captured during time 2 is p_2. To have a capture history of 11, *both* of these events have to occur (i.e., captured in each period), so we multiply the two probabilities together.

Similarly, to have a history of 01, an animal must avoid capture in time 1 and be captured in time 2. Thus, the probability is: $q_1 * p_2$. And, the opposite is true for a history of 10: $p_1 * q_2$.

These are the probability statements that a maximum likelihood estimator will use to find the best estimate for p. And, when we estimate capture probability, we can estimate N. Remember that, in general form, $N = n/p$? That is, if we can estimate the proportion of the population that we didn't capture, we can estimate population size—because we know the number of animals we captured. In the table above, if you add the first three cells together, we can calculate how many animals we captured as the sum of animals with 11, 10, and 01 capture histories. Or, mathematically, $r = (n_1-m_2)+(n_2-m_2)+m_2$.

Chapman wasn't satisfied

Like all maximum likelihood estimators, the L-P estimator is unbiased for very large samples. But, it is biased for small samples. This bias can be shown by putting 10 white balls in a hat (a very small population, which ensures small samples). If you randomly draw a sample of 4 balls (a small sample) and mark them, we'll now have 6 unmarked and 4 marked balls in our 'population' of balls. We know (because we set up this experiment) that the marked proportion of the population is 0.4. And, $n_1 = 4$.

Now, if we draw a sample of four more balls, we could get a variety of results. Using simple logic, if 40% of our population is marked, we might predict that we'll most likely get two marked balls in our sample of four balls that we randomly draw. You'll notice that it is impossible for us to get exactly 40% of our sample (0.4 * 4 = 1.6, and we are not splitting balls in half…just as you cannot split animals in half during sampling!). And, that is the reason that the L-P is biased for small samples.

To carry the example forward, if $m_2=2$ and $n_2=4$, then our population estimate would be: $\hat{N}=4*4/2 = 8$.

Now, let's allow a random occurrence to happen…and let's say we got, instead, 1 marked ball in our sample. Such an event might occur with such a small sample. Now, $m_2 = 1$. And, $\hat{N} = 4*4/1 = 16$. A change in one marked ball in our sample doubles our population estimate.

We can pause and state that most scientists would never attempt to do a mark-recapture exercise if the size of the population was thought to be near 10! We have used such a small population and small samples to illustrate the problem. But, it might be common for your samples to be less than 30 in some situations. For example, what happens to the L-P method if $m_2 = 0$ (no marked animals captured)? $\hat{N}$ is undefined, as you cannot divide n_1*n_2 by 0.

The Lincoln-Petersen estimator was modified by Chapman, and the resulting estimator is known as the "**Chapman modification**" (yes, we biologists are often very literal in the names we give to methods!). The modification has less bias for small samples than the L-P estimator. Chapman based his modification on the hypergeometric probability distribution, which is used when sampling without replacement. We mention this simply to note that there is statistical rigor and theory behind the modification. However, it is not a maximum likelihood estimator:

$$\hat{N} = \frac{(n_1 + 1) \times (n_2 + 1)}{(m_2 + 1)} - 1$$

The modified method does solve, mathematically, the problem of situations when m_2 is 0. However, we would remind the reader that if a study does not result in capturing a marked animal, it is more likely that the best course of action is to reconsider the study design. How might you mark more animals during time 1? Or, how might one might work with a species that is either so rare or so trap-timid that no marked animals are caught.

You try it!

Estimate $\hat{N}$ using the Pollock et al. (1990) quail data from above. Were the samples small enough to warrant use of the Chapman method? Compare your answers to those found at the end of the chapter.

Schnabel had more data...

The **Schnabel method** expanded the basic philosophy of Lincoln-Petersen to be used when the number of capture occasions is larger than 2. When Zoe Emily Schnabel, a mathematician at the University of Wisconsin, developed this method in the 1930's, only 2-occasion studies could be used. So, her work redefined possibilities for wildlife science. The Schnabel method is a closed form method, which means it can be condensed to an equation. But, it is an approximation to a maximum likelihood estimation method that can be found in modern software tools. White et al. (1982) noted that many biologists have disregarded a note that Schnabel wrote about her method: *"It should be emphasized, however, that none of the solutions can be expected to provide more than a general estimate of the order of magnitude of the total population."*

For this method, C_t is defined as the number of captures during time t. R_t is the number of recaptures (marked animals, captured again) captured during time t. And, M_t is the number of animals marked in the population, at time t.

$$\hat{N} = \frac{\sum_t (C_t \times M_t)}{\sum_t R_t}$$

Note: To properly calculate M_t, *do NOT count animals that are marked for the first time during time t. That is,* M_t *is the number of animals marked in the population just as you begin to conduct your observations of the traps or nets. Or, you can also think of* M_t *as the number of marked animals in the population that were at 'risk' of being trapped during time t.*

You try it!

During 2006, students from the University of Nebraska-Lincoln traveled to a 7-ha island off the coast of Puerto Rico. The island was named Isla Magueyes, and it had a population of Cuban rock iguanas (*Cyclura nubila*). These iguanas are a species of conservation concern elsewhere in their range, but their population was causing damage on Isla Magueyes. A small population of iguanas had been left on the island when a zoo was removed. No one knew how many iguanas were on the island.

So, the students marked animals each morning for three days. On the first day, the students found 155 iguanas, and they squirted them with small marks of latex paint to mark them. The marks were not individual ID's, but the marks were easy to see on day 2 and 3.

On day 2, the students saw 109 marked animals and 66 unmarked animals. They marked any animals that were not marked. On day 3, the students observed 116 marked animals, and only 15 unmarked animals. Use the Schnabel method to estimate the population size.

	Animals Observed	Animals with Marks When Seen
Day 1:	155	0
Day 2:	175	109
Day 3:	131	116

The first step is to complete a summary table. M_t can be a tricky statistic to calculate, so we have added a 'newly marked column' here. The numbers of newly marked animals have been added to help you stay on track. In addition, we have noted that $M_1 = 0$ (no marked animals in the population before the first period).

Day	C_t (Captures during time t)	R_t (Number of recaptures captured during time t)	Number newly marked	M_t (Number of animals marked in the population, just before you conduct trapping during time t)
1			155	0
2			66	
3			15	

Population size estimate: _____

See the end of this chapter for answers after you have given this a try.

Modern closed-population methods

The analytical methods described above can be accomplished by **batch marking** individuals, rather than going to the effort of applying an individual identifying letter and/or number combination to each animal. For example, the Lincoln-Petersen and the Schnabel data can be gathered by applying a basic mark—like the splotch of paint used on the iguanas in our example. In the third time period, there is no need to know whether a marked animal was previously marked in the first or second time period—it is enough to know that it was marked at some point in the past.

The assumption of the Lincoln-Petersen and the Schnabel is that all animals in the population at a given time period have the same probability of capture. What if that is not true? For example, what if the iguanas that encountered students were less likely to be in the accessible areas of the island on the second day, while their unmarked friends were content to sunbathe on sidewalks that would eventually be used by students who marked them? If we break that assumption, the L-P estimator does not work as it should.

The alternative is to use **individual marks**, which allow the application of more modern closed-population estimation methods (Figure 8.2).

Figure 8.2: *A prairie rattlesnake (Crotalus viridis) is marked with a Passive Integrated Transponder (PIT) tag to provide individual identification upon recapture. Photo provided by Dennis Ferraro and used with permission.*

Otis et al. (1978) proposed several models for use in estimation of population size that are still in general use today. The most important contribution was the notion that animals could become "trap happy" (recapture rates are higher than initial capture rates) or "trap shy" (recapture rates are lower than initial capture rates). Today, we define **capture rate**, p, as the probability of initial capture and **recapture rate**, c, as the probability of recapture after being captured at least once.

And, although Otis et al. (1978) did not use AIC to compare alternative models that described variation in capture probability, their suggestion to compare multiple models led biologists toward the idea that capture estimates from "poor" models should not be used. Rather, estimates from models judged to be better should be used. The application of AIC for model selection (see Chapter 4) in ecology developed from this work.

By the use of individual marks, we can develop individual capture histories, which are not possible with batch marking (e.g., data often used for Schnabel method). Capture histories are important, because we can tie specific probability statements to each capture history—that is, each capture history has a specific probability of occurrence. For example, the probability of an animal having specific capture histories in a 10-occasion, closed-population, mark-recapture study is:

Capture history	**Probability**
0100111000	$(1-p_1)p_2(1-c_3)(1-c_4)c_5c_6c_7(1-c_8)(1-c_9)(1-c_{10})$
1000110101	$p_1(1-c_2)(1-c_3)(1-c_4)c_5c_6(1-c_7)c_8(1-c_9)c_{10}$
0000001101	$(1-p_1)(1-p_2)(1-p_3)(1-p_4)(1-p_5)(1-p_6)c_7c_8(1-c_9)c_{10}$

With an adequate sample of marked animals, you may use software that uses maximum likelihood estimators to estimate the capture and recapture rates, as well as population size for the species in your study. In the above example, we show variation caused by behavior (capture and recapture rates not equal), as well as time (different capture and recapture rates for each time period). However, you might explore alternative models. For example, we might hypothesize that behavior has no impact on recapture rates, and hence $c = p$. Therefore, the probability statement for the first capture history above could be represented using only p and no c (the recapture probability):

0100111000 $\quad\quad (1-p_1)p_2(1-p_3)(1-p_4)p_5p_6p_7(1-p_8)(1-p_9)(1-p_{10})$

A third model that we might explore could state that behavior causes differences in capture and recapture, but time is not important. So, the probability statement for the first capture history in the table above would not have any time-specific references (no subscript numbers):

0100111000 $\quad\quad (1-p)p(1-c)(1-c)ccc(1-c)(1-c)(1-c)$

And, last (at least for our simple example), we might hypothesize one last model—a null model—that neither behavior nor time cause capture rate to vary. Thus, the likelihood statement can be represented using only p and not c, the recapture probability. It also does not include time-specific p's (no subscripts):

0100111000 $\quad\quad (1-p)p(1-p)(1-p)ppp(1-p)(1-p)(1-p)$

Which model is best? We could use **AIC** to tell us which model is the most likely to represent reality, given our data (Chapter 4). And, we would report the population size and capture probabilities estimated by the best model.

Lukacs (2014) provides a description of many modifications of the closed-population mark-recapture models. We suggest the beginning user start with **full likelihood** (p and c) or **Huggins'** (p and c) models. Each allows the user to explore the basic models we have outlined above.

Conclusion

Simple models to estimate population size require the assumption of closure during the sampling period: no births, deaths, or movements in or out. The history of estimating the size of a population started with a simple 2-occasion mark-recapture experiment: the Lincoln-Petersen scenario. The Schnabel method allows an estimate based on more than one period, which paved the way for modern methods. All of these methods are based on the same general concept: for a set of captured and marked animals, if you have a second capture occasion and you look at the second set of captured animals, the proportion of the animals with marks in your sample is assumed to be the same proportion of the population that you marked during the first capture period. Most modern methods require individual marks on animals, and the methods allow the exploration of multiple hypotheses regarding the variation of capture probability in the population during the study.

Answers: You try it!

Pollock et al. (1990) quail problem: 148 northern bobwhite (quail) were marked with leg bands; of 82 shot quail in the next month, 39 were marked with leg bands.

$n_1 = 148$
$n_2 = 82$
$m_2 = 39$

$\hat{N} = 311$ (*rounded*)

Chapman modification for the Pollock et al. (1990) quail problem:

$\hat{N} = 308$ (*rounded*).

Were the samples small enough to warrant use of the Chapman method? The difference between 308 and 311 (a measure of the bias of the Lincoln-Petersen estimator) is pretty small (just a difference of 3 animals).

Schnabel method (iguana problem):

	Animals Observed	Animals with Marks When Seen
Day 1:	155	0
Day 2:	175	109
Day 3:	131	116

Day	C_t	R_t	Number newly marked	M_t
1	155	0	155	0
2	175	109	66	155
3	131	116	15	155+66=221

$$\hat{N} = \frac{(155 \times 0) + (175 \times 155) + (131 \times 221)}{0 + 109 + 116} = \frac{0 + 27125 + 28951}{225} = 249.2$$

References

Laplace, P. S. 1783. Sur les naissances, les mariages et les morts. Mémoires de l'Académie Royale des Sciences de Paris 693–702.

Lincoln, F. C. 1930. Calculating waterfowl abundance on the basis of banding returns. United States Department of Agriculture Circular 118: 1–4.

Lukacs, P. 2014. Chapter 14: Closed population capture-recapture models. *In* Program MARK: a gentle introduction, 12th edition, Cooch, E. and G. White, eds. Online: http://www.phidot.org/software/mark/docs/book/pdf/chap14.pdf

Petersen, C. G. J. 1896. The yearly immigration of young plaice into the Limfjord from the German Sea. Report of the Danish Biological Station (1895) 6: 5–84.

Pollock, K. H., J. D. Nichols, C. Brownie, and J. E. Hines. 1990. Statistical inference for capture-recapture experiments. Wildlife Monographs 107: 3-97.

White, G. C., D. R. Anderson, K. P. Burnham, and D. L. Otis. 1982. Capture-recapture and removal methods for sampling closed populations. Los Alamos National Laboratory. LA-8787-NERP. 235 pp.

For more information on topics in this chapter

Amstrup, S. C., T. L. McDonald, and B. F. J. Manly. 2005. Handbook of capture-recapture analysis. Princeton Univ. Press: Princeton, NJ.

Conroy, M. J., and J. P. Carroll. 2009. Quantitative conservation of vertebrates. Wiley-Blackwell: Sussex, UK.

Williams, B. K., J. D. Nichols, and M. J. Conroy. 2002. Analysis and management of animal populations. Academic Press, San Diego.

Citing this primer

Powell, L. A., and G. A. Gale. 2015. Estimation of parameters for animal populations: a primer for the rest of us. Caught Napping Publications: Lincoln, NE.

A maned wolf (Chrysocyon brachyurus) is measured by a research team prior to marking with a radio-telemetry transmitter in Brazil. Photo taken by Adriano Gambarini. Lucia Corral: left, Nucharin Songsasen: right) and used with permission of the photographer.

Ch. 9

Known fate[1]

"You shall know the truth, and the truth shall make you mad."
-- Aldous Huxley

"I am always anxious to know what has happened while I've been asleep."
-- Christian Lacroix

Questions to ponder:
- *How does known-fate data differ from other types of data used to estimate survival?*
- *What is the difference between left- and right-censoring?*
- *How do I calculate nest survival with the Mayfield method?*
- *How is "staggered entry" handled with Kaplan-Meier methods?*

Known-fate data

We will start to assess ways to estimate survival by starting with the simplest analytical method—known-fate data. We note that these data are not necessarily the easiest to collect, nor are they the cheapest to collect. But, the structure of the models is the simplest.

Known-fate is a type of data that involves the marking of a cohort of individuals that is re-visited at a later time to determine their fates. The unique quality of known-fate information is that we have 100% certainty, theoretically, of finding the individuals and assessing their fate. In Chapter 8, we discussed encounter probabilities—a necessary "evil" in parameter estimation, because we cannot typically guarantee that we will catch, trap, see, or otherwise encounter a specific individual during a specific time period. However, this problem disappears for a select group of situations for which we can provide known-fate information.

What kind of methods or situations provide known-fate data? The use of radio-telemetry is a grand example. Because we place a radio or satellite tag that allows the tracking of an animal's location (and in some cases, the tag even broadcasts a prediction of the animal's live/dead status), we can find the animal during any time period we wish—providing that the tag is still working and the animal has not escaped the geographic breadth of our relocation capabilities. The monitoring of plants and nests are other examples of known-fate scenarios. Plants can be

---

[1] *With thanks for content to Michael Conroy, Jay Rotella, Evan Cooch, William Clark, and Gary White*

labeled with a flag and relocated with ease. Nests do not move, unlike their creators. So, we can mark the position of nests and find them in the future.

Known-fate approaches can also be used to study dynamics other than survival. In Chapter 3, we used a tag loss study to introduce the concept of a maximum likelihood estimator. In that example we examined the probability that fish would lose their tags. The biologists put 10 fish in a tank, and each fish is marked with the special tag. After one day, the animals are checked. The biologists estimated the probability of tag retention over the 1-day period of time. The tanks of fish could be relocated and their 'fate' (tagged or not tagged) assessed perfectly. We could use known-fate approaches to estimate the probability of a specific behavior of animals under observation, where we consider the presence of behavior or lack of a behavior to be a binomial success/failure.

We can also pause and note that ecologists are not the only ones who use known fate information to estimate survival. Biologists who engage in medical studies can typically track patients and relocate them throughout a study to determine the success or failure of a certain treatment. Mechanical engineers can track specific products—like washing machines and airplanes—to estimate longevity. There is no need to estimate encounter rates, because the subject can usually be located with 100% probability. Of course, things can go wrong with tagged animals, and we discuss this below!

The ability for near-perfect relocation of a study subject simplifies our analysis. In Chapter 10, we will discuss the estimation of survival when we cannot relocate our subjects during each time interval—using a method established by Cormack, Jolly, and Seber (CJS). Here, we see the simplicity of known-fate analyses, for which we do not need an encounter rate (p). Given capture of an animal in occasion 1, the probability statements for the subsequent observations (enumerated by the capture histories using a simple coding of "1" for "survived time period" and "0" for "died during time period") can be written as follows:

Animal	Capture history	Probability statement
Animal #1	111	$S_1 \cdot S_2$
Animal #2	110	$S_1 \cdot (1-S_2)$

Thus, our known-fate information tells us that Animal #1 was alive (1) in all three time periods, so the probability of observing the animal being alive from occasion 1 to occasion 3 is simply the product of two survival probabilities. Animal #2 is known to have died (0) after being observed alive at the beginning of occasion 2, so the probability of observing this is the product of one survival probability, S, and one mortality probability $(1-S)$.

In Chapter 3, we developed a maximum likelihood estimator for known fate data that can be used to estimate survival with n trials and y successes:

$$\hat{S} = \frac{y}{n}$$

And, in Chapter 4, we used a known-fate example with northern bobwhite (quail [*Colinus virginianus*]), and we estimated sex-specific survival estimates for a given time period for the following scenario:

> *100 radio-marked quail*
> *50 males, 50 females*

At end of time period:
> *20 males alive*
> *35 females alive*

The simple period-specific estimate for S_{males} would be

$$S_{males} = 20/50 = 0.40$$

The simple period-specific estimate for $S_{females}$ would be

$$S_{females} = 35/50 = 0.70$$

The known-fate approach for these data has some underlying **assumptions**:
- The individuals in our sample are independent
- Capture probability, with our tagging method, is 100%
- All individuals have the same underlying survival probability when individuals are modeled with the same survival parameter (e.g., if a "male" survival parameter is estimated for all males, all males are assume to have the same probability of survival)

The latter assumption simply indicates that our method is to sample a group of individuals, and we use the combined patterns of fates of all sampled individuals (e.g., all males in the example mentioned above) to estimate a probability of survival that is associated with every individual in the sampled group. In our example, we've divided quail into males and females, so the survival estimate that we receive for males applies to all males. We could also pool males and females to estimate one survival probability that would apply to all quail. And—if we had information available—we could divide males and females into sub-groups of juveniles and adults to estimate survival of adult males, juvenile males, adult females, and juvenile males. The only method we have to estimate a survival rate for a specific individual is to use individual covariates to predict survival based on some characteristic of the individual—for example, as we saw in Chapter 6, we could also collect mass information for each bird in our sample, and we could model survival as a function of mass.

Mayfield's known-fate model for nests

The history of known-fate models is a good way to learn about the method, as we typically start with simple structures and move towards more complex structures as methods develop over time. Harold Mayfield made an early advance in known-fate models for ecology in 1961 when he published methods for the estimation of the daily probability of survival for avian nests (Mayfield 1961). Of course, this method can be, and has been, generalized to estimate daily survival of animals and plants. We start with Mayfield's method because of its relative simplicity, and because it showcases an ecologist thinking about his data—and adjusting his analysis to reduce bias that he perceived in the standard analytical methods of his time.

The context of Mayfield's work (for those of you who do not spend countless hours searching for bird nests) is that avian biologists have to search pretty diligently to find bird nests—it's similar to trying to trap a species of animal that is fairly difficult to trap. Eggs are laid and incubation begins, but the field biologist might not find the nest for a few days. Of course, once the nest is found, the biologist can return again and again to monitor the nest in known-fate fashion.

The data available to the biologist is the first date of observation (t_0), and the last day that the nest was observed. We also have the status on that last date (was the nest successful, or did it fail). In certain cases, we might have to "censor" (more on this later in the chapter) our data, because the status was unknown after a certain censor point (perhaps a technician forgot to check the nest, perhaps the nest could not be found again, or perhaps the research team had to end the study before the nest's cycle was complete, etc.).

Mayfield's methods were also based on the context that avian biologists typically check nests over periods of days at 2-, 3-, or 4- day intervals. Nests are typically not checked every day because daily checks could cause too much disturbance at the nest. In addition, visiting rates may vary among nests (2-day intervals for one nest, and 4-day intervals for another nest that is logistically harder to reach). Still at each visit, the fate of the nest (still 'alive' or failed) is recorded. And, there is an end to the monitoring, the date for which varies depending on when the nest fails or a brood disperses from the nest.

Prior to Mayfield's work, waterfowl and songbird biologists had simply reported the raw success of nests as a percentage of nests that were successful. So, no matter when a nest was found (early in its cycle or almost before completion), the nest was included in the sample of "nests at risk". And the simple maximum likelihood estimator was used to estimate "nest success": the proportion of nests that succeeded.

$$\hat{p} = \frac{y}{n} = \frac{successes}{trials}$$

Mayfield's contribution

Mayfield (1961) advanced the field of known-fate analyses, because he acknowledged that failed and successful nests are not usually found by biologists with the same probability. To clarify this, we can summarize:
- Most nests are found after egg-laying has begun
- Normally, only active nests (with attending parent(s)) are found
- We often miss early failures that happen before we find some nests

The last point was Mayfield's main concern. Because of the lag in finding nests, many failed nests are probably never incorporated into the sample of nests—that is, a nest that was found on day 10 had to, by definition of the manner in which nests are found, survive until day 10. If the nest had failed on day 8 (never found or found after it failed), it would have never been entered into the sample. Thus, "apparent nest success" (our modern term for the proportion of nests that succeed) is biased high, relative to actual nesting success.

Rotella (2014) provides a convincing analysis of the potential bias in nest survival estimates that occurs because of the lag in finding nests after the eggs are laid. If all nests are found on day 0 (which can happen if the females are radio-marked or if the biologist is working in a colonial species where nests are constantly monitored in a small area), there is no bias. But, if nests are not found, on average, until day 10, nest success is biased by 12%. That is, an "apparent nest success" estimate of 35% should actually be adjusted to 23%--because nests that failed before day 10 were not included in the sample. Rotella's (2014) bias estimate is based on a 35-day nesting scenario, but similar biases were Mayfield's concern as well—for nesting intervals of any length.

Mayfield's solution was to avoid using the nest as the "trial". Instead, Mayfield proposed using a risk-day as the trial. Mayfield (1961) theorized that all nests have the same probability of failing, and he treated all nests under observation as the sample. Therefore, Mayfield's advancement was to estimate "daily nest survival"—instead of "nest survival probability". The method accounts for the fact that some nests were not observed for the entire nest period.

Mayfield's estimator for daily nest survival is:

$$\hat{S}_d = 1 - \frac{number\ deaths}{Exposure\ days}$$

Daily survival is estimated using the exposure days of all nests as the number of trials, and the mortality rate (deaths per exposure days) is changed to a survival rate. And, after the daily nest survival is estimated, the probability of a nest (with a nest interval of L days) can be estimated as

$$\hat{S}_p = \hat{S}_d^L$$

As we noted earlier, situations do arise (usually unexpectedly) for which **censoring** is necessary. Nests that cannot be found or monitored again should be removed from the exposure day count. However, we don't remove all of the data from these nests—the nest was present and monitored until a certain date, so the information that we know remains a part of our sample.

As an example, if we used Mayfield's simple estimator to estimate survival of elk (*Cervus canadensis*) in a sample that required censoring, we would now use this adjusted formula:

Figure 9.1: *Male elk (USFWS public domain photo by Paul Prado).*

$$\hat{S}_d = 1 - \frac{number\ deaths}{Exposure\ days - number\ censored\ animals}$$

If a given elk was in the sample for 15 days, but could not be found (perhaps radio failure, perhaps moved off the study area) on day 16, we would end its observations. Although there are 16 exposure days, one of the exposure days is removed because we censored the animal on day 16. When that elk's data are added into the sample for use in the equation above, we would include 16 exposure days and one censored animal (Figure 9.2; results in a net of 15 exposure days: 16-1 = 15) and no deaths. Exposure days, censor events, and deaths from all elk in our sample would be used to estimate S_d.

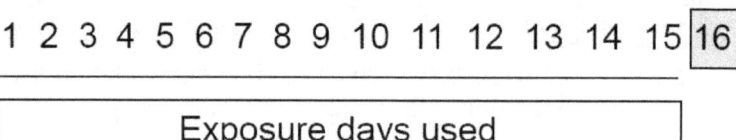

Figure 9.2: *Calculation of exposure days for animal "lost" on day 16. Exposure days are limited to the days for which the fate of the animal, at the end of the time period, is known.*

Mayfield's methods have the following assumptions and characteristics:
- All nests (or more generally, "subjects") have constant survival over the nest interval, L
- Subjects are not required to enter the study at the same time (known as "staggered entry" to the sample)
- Survival is constant for the entire sample; thus, in staggered entry, survival is not different for the newly added individuals
- The estimate for variance is based on number of exposure days (number of trials), so studies with fewer exposure days will have a less precise estimate of survival.

The variance (and remember, $SE = \sqrt{var}$) for the estimate of daily survival ($\hat{S}_d$) from the Mayfield method, with exposure days, d_i, for n nests, is (Skalski et al. 2005):

$$\hat{var}(\hat{S}_d) = \frac{\hat{S}_d(1-\hat{S}_d)}{\sum_{i=1}^{n} d_i}$$

And, using the delta method (see Chapter 5), the variance for the period-specific (over L days) probability of nest survival is:

$$\hat{var}(\hat{S}_p) = [L \cdot \hat{S}_d^{L-1}]^2 \cdot \hat{var}(\hat{S}_d)$$

A problem encountered by Mayfield was what to do with the last interval in which a failure was observed. Again, the context is that a biologist would observe an active nest on a certain day and then check the nest again, perhaps 4 days later. If a failure was observed, it was obvious that it happened sometime in the last 4 days…but when? This uncertainty affected the exposure days—because it should be obvious that for a large sample of nests which fail, they do not always fail on the 4th day of a 4-day interval. Presumably, some fail on the 1st day of the interval, some on the 2nd, some on the 3rd, and some on the 4th.

Mayfield postulated (which has been subsequently supported by Manolis et al. 2000), that if we assume constant survival, the average date of failure in a large sample of nests would be the midpoint of the last interval. So, in our example, we would use two exposure days rather than four.

Mayfield example: Georgia wood thrushes

Figure 9.3: *An adult wood thrush. Public domain photo.*

To provide a simple Mayfield example, let's look at some real nests of wood thrushes (*Hylocichla mustelina*, a medium-sized songbird) in Georgia in 1996 (Lang et al. 2002). Wood thrush nests were found during months of April, May, June, and July. Lang and his associates recorded the date on which each nest was found, the last date the nest was known to be active, the last day the nest was checked, and the nest's fate as of the last check. Dates were recorded as day-of-season, where April 1 = day 1 and therefore May 1 = day 31, which makes calculations much easier!

We can use the table of data (below) to calculate the exposure days and the number of failures needed to estimate daily nest survival. We first calculate the number of exposure days during times that are known to be within active intervals for each nest. Then, we look at the last interval for nests that failed, and calculate (using the Mayfield approach) the number of exposure days to use during that interval (1/2 of the interval). We then tally up the number of failed nests.

The apparent nest success estimate for this sample of 13 nests (of which 5 were successful) is $5/13 = 0.385$. So, 39% of our nests survived. But, many of them were not found on the first day of the nesting cycle. Finding nests is difficult…

We can use Mayfield's method to estimate daily survival as:

$$\hat{S}_d = 1 - \frac{number\ deaths}{Exposure\ days} \qquad \hat{S}_d = 1 - \frac{8}{176} = 1 - 0.0455 = 0.9545$$

And, variance:

$$\hat{var}(\hat{S}_d) = \frac{\hat{S}_d(1-\hat{S}_d)}{\sum_{i=1}^{n} d_i}$$

$$\hat{var}(\hat{S}_d) = \frac{0.9545(1-0.0455)}{176} = \frac{0.9111}{176} = 0.0052$$

Thus:

$$SE(\hat{S}_d) = \sqrt{\hat{var}(\hat{S}_d)} = \sqrt{0.0052} = 0.0719$$

Day* found	Day last active	Exposure days during active intervals	Day of last check	Days assumed alive during failure interval**	Total Exposure Days	End fate	Failure
67	83	16	85	1	17	F	1
11	11	0	16	3	3	F	1
61	78	17	81	2	19	F	1
86	86	0	89	2	2	F	1
32	40	8	43	2	10	F	1
40	47	7	47		7	S	0
72	96	24	96		24	S	0
46	62	16	65	2	18	F	1
53	72	19	72		19	S	0
82	87	5	90	2	7	F	1
93	99	6	101	1	7	F	1
51	71	20	71		20	S	0
50	73	23	73		23	S	0
TOTALS:		161		15	176		8

*Day-of-season: April 1 = day 1; **Using the Mayfield approach to only include half of the exposure days during the last interval when the interval ends in a failure.

The SE of 0.07 indicates a fairly high level of uncertainty around the daily survival estimate (a 95% confidence interval would be approximately ±0.14). But, remember that this is a small sub-sample of 13 nests that we are using for our example. We have only 176 exposure days, and you should see by looking at the equation above (for the variance of the daily survival) that if we had the same estimate of daily nest survival ($\hat{S}_d$) but 1000 exposure days, our variance estimate would be much, much lower.

What is our Mayfield-corrected probability of nest survival? Wood thrushes have, on average, 24-day nest periods (12 days incubation and 12 days nestling care) before the nestlings fledge. To extrapolate the daily survival to 24-day survival to estimate the probability that a wood thrush nest produces fledglings, we can use:

$$\hat{S}_p = \hat{S}_d^L \qquad \hat{S}_{24} = \hat{S}_d^{24} = 0.9545^{24} = 0.3271$$

And, the variance as:

$$\hat{var}(\hat{S}_p) = [L \cdot \hat{S}_d^{L-1}]^2 \cdot \hat{var}(\hat{S}_d)$$

$$\hat{var}(\hat{S}_{24}) = [24 \cdot 0.9545^{23}]^2 \cdot 0.0052 = 0.1248$$

Thus, the SE:

$$SE(\hat{S}_{24}) = \sqrt{\hat{var}(\hat{S}_{24})} = \sqrt{0.1248} = 0.3533$$

So, our potentially-biased apparent nest success estimate was 39%, and our Mayfield adjusted estimate was approximately 33%. This sample shows the bias that Mayfield tried to correct with his method—Lang and his colleagues, like most avian ecologists, apparently missed some early nest failures that were not included in the naïve estimate of apparent nest success. Mayfield's estimator gave us a tool to eliminate this bias.

Significant digits

Before moving on, we pause to emphasize the value of using at least 4 decimal places when reporting and manipulating survival probabilities. If we were to report fewer decimal places, our 24-day estimate could vary, because of rounding. Consider similar factors of raising daily survival probabilities of adults, for example, to annual probabilities by raising to the 365th power---similar effects of rounding would definitely occur. This effect could lead to significant biases in simulation models.

Below, the daily survival estimate from the wood thrush example above is provided with 4 decimal places and fewer, prior to estimating the 24-day nest survival probability. The difference in the predicted nest survival is about 3.5%, using the same initial daily survival estimate!

$$0.9545^{24} = 0.3271$$
$$0.955^{24} = 0.3312$$
$$0.95^{24} = 0.2920$$

Kaplan-Meier methods

Three years before Mayfield's (1961) publication, Kaplan and Meier (1958) provided a survival estimator that has also been incredibly influential to known fate survival estimation. The method provided, which has become known as the Kaplan-Meier method (again, we pause to note that one method to achieve immortality is to develop a new analytical method that will most likely be named after you!), sought to combat two problems with the standard, very simple approaches to survival estimation in known-fate situations, where

$$\hat{S} = \frac{y}{n}$$

Kaplan and Meier (1958) provide the problem statement for their work:
> "*In lifetesting, medical follow-up, and other fields the observation of the time of occurrence of the event of interest (called a <u>death</u>) may be prevented for some of the items of the sample by the previous occurrence of some other event (called a <u>loss</u>). Losses may be either accidental or controlled, the latter resulting from a decision to terminate certain observations. In either case it is usually assumed in this paper that the lifetime (age at death) is independent of the potential loss time; in practice this assumption deserves careful scrutiny. Despite the resulting incompleteness of the data, it is desired to estimate the proportion P(t) of items in the population whose lifetimes would exceed t (in the absence of such losses), without making any assumption about the form of the function P(t).*"

Very simply, Kaplan and Meier were worried that:
- Mortality was not constant through life, and in fact they wanted to model the survival function, $P(t)$, to help with insurance estimation and medical evaluations of humans.
- Often, individuals are removed from the sample (they refer to this as *loss*, we would refer to this as **censoring**), making the simple $S=y/n$ formula inadequate.

The context, then, of a Kaplan-Meier analysis of survival is one in which the study can be broken into intervals (regular or irregular), each with a constant risk of mortality within the interval. Such studies are used extensively in medical, engineering, and product reliability studies and are often referred to as "time to failure" studies. Ecologists soon warmed to this technique, because it allowed graphical representation of variation in survival through time. Remember that the Mayfield method assumes constant survival, which does not easily allow assessment of potential variation in survival during the nesting period.

Kaplan and Meier's model structure is as follows:

$$\hat{S}_t = \prod_{i=1}^{t}\left(\frac{n_i - d_i}{n_i}\right)$$

where
- $S(t)$: probability of an animal surviving t units of time from the beginning of the study
- We know, for each animal, the first and last date of observation and the fate of the animal at each time step
- n_i = number of animals alive and at risk in time period I
- d_i = number of deaths during time period i

Note that we can see that survival is a product (because of the use of $\prod$, which directs us to take the product) of sub-units in which survival is calculated as in Mayfield. In fact, the parenthetical portion of the formula is exactly the same as Mayfield's estimator for survival. The variance for survival in Kaplan-Meier methodology is based on the number of animals (*not exposure days as in Mayfield*).

Kaplan and Meier labeled this estimate of S_t as the **product limit estimate**, which they envisioned as a function that described the cumulative effects of period-specific mortality on a population. If the survival probability for an individual time period, s_t, could be determined, then the survival probability for all time periods, S_t, could be estimated as the product of the survival for the individual time periods.

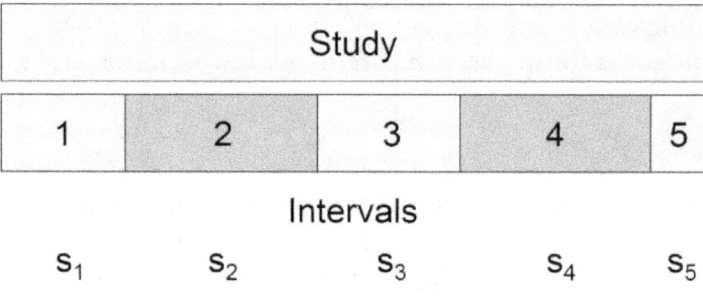

Figure 9.4: *Temporal depiction of a study with the goal to estimate the probability of survival for the entire study period (S_t). The study has 5 time periods, each with an estimate of survival (s_t). S_t, then, is the product of all s_t's.*

We note that, for the Kaplain-Meier product limit estimate, the time periods do not have to be of the same duration. As an example, consider data approximated from Fries (2002) for human survival in the United States in 1900. We can look at time periods of interest: early survival during ages 0-10, a period of low mortality from 11-50, and periods of increasingly higher mortality from 51-70, 71-80, and 80-100.

As the following table shows, we can estimate the period-specific probability of surviving any of the intervals. In 1900, in the United States, young children had a 90% probability of surviving until age 1. And, that 1-year survival probability was almost equal to the 9-year probability of surviving from age 1 to age 10 (S = 0.8889). At age 70, people only had a 17% probability of being alive at age 90. And, there was a 0% probability of surviving to 100, for those who were 90.

Age Interval (t)	No. at risk	No. deaths	s(t)	S(t)
0-1	100	10	0.9000	0.9000
1-10	90	10	0.8889	0.8000
10-50	80	20	0.7500	0.6000
50-70	60	30	0.5000	0.3000
70-90	30	25	0.1667	0.0500
90-100	5	5	0.0000	0.0000
100+	0			

The cumulative survival, S_t, can be calculated at any point, and these estimates are then shown in the figure. A graphic advantage to Kaplan-Meier-type survival curves is the ease with which one can identify periods of high risk and low risk for a population over time (Figure 9.5).

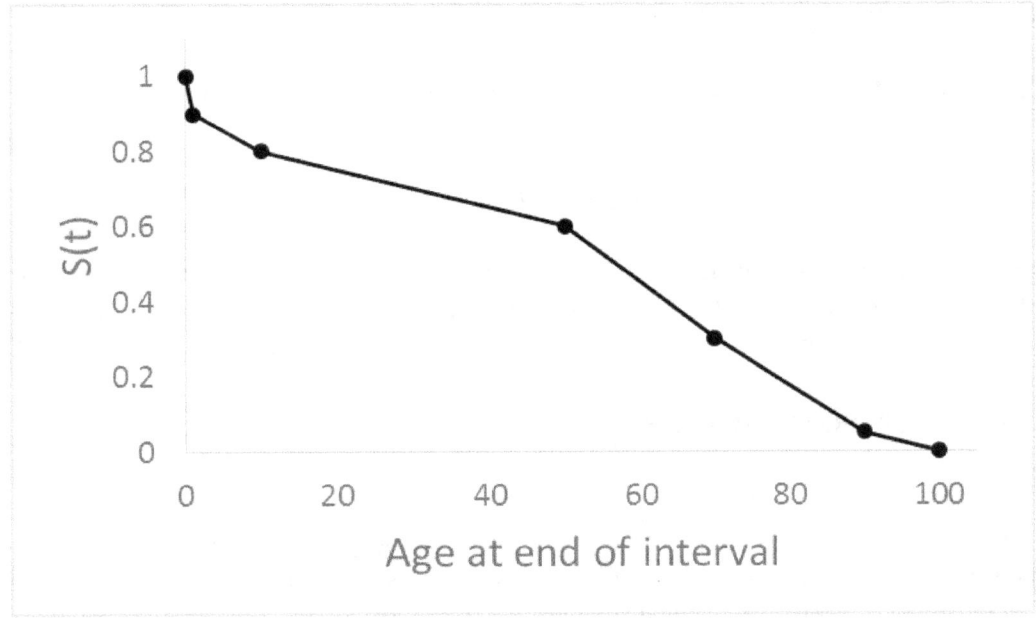

Figure 9.5: *A Kaplan-Meier estimate of cumulative survival, S_t, over 100 years of life, for humans in the United States.*

Kaplan and Meier's example

It may be instructive to look at the example that Kaplan and Meier provided to show the merits of their product-limit estimator. The context is a two-year evaluation of a set of individuals that are being monitored for survival during the first two years of life. Two trials, or samples, are available to provide survival information: the first is a small trial for two years, and the second is a larger trial that only lasted for one year. Both trials provide information about survival during year-one. But, only the first trial provides year-two information.

Observations/Data*	Samples/Trials	
	I	II
Initial cohort	100	1000
Deaths in first year	70	750
Deaths in second year	15	--
One-year survivors	30	250
Two-year survivors	15	--

From Kaplan-Meier (1958)

As Kaplan and Meier (1958) describe, it is easy to look at Sample I and determine that S_2 can be estimated as 15/100 = 0.15 (a 15% probability of surviving for two years). But, this ignores the 1000 samples tested during the second study. Unfortunately, we cannot lump the two studies together to estimate S_2 as 15/1100, because Sample II did not follow individuals for year 2.

But, Kaplan and Meier proposed that information from both samples could be used to inform the year-one survival estimate, or s_1. Thus, if we combine the data from both samples, we can estimate s_1 as (30+250)/(100+1000) = 0.255. And, we note that this is different estimate for s_1 than if we only used Sample I (30/100=0.300).

Now, Kaplan and Meier completed the calculations by using Sample I's data (the only available) for year-two survival, s_2 = 15/30 = 0.500.

And, the estimate for S_2 can now be obtained as 0.255*0.500=0.127.

This result, we note, is different than if we only used Study I, and Kaplan and Meier (1958) argued that our current estimate for S_2 is now better-informed because we have used all available data. That is how the Kaplan-Meier method was introduced to the world.

Kaplan-Meier assumptions

The **assumptions** of the Kaplan-Meier method are:
- An animal is considered at risk if alive at start of time period (see example below)
- Survival is assumed constant for all animals during specific time periods, but there is no assumption of constant survival throughout the study.
- K-M methods do not require all animals enter at the same time

- Newly tagged animals (in staggered entry) are assumed to have the same survival function as previously tagged animals
- Animals represent a random sample of the population
- Animals are independent
- Working radio-tags are always located, when using radio-telemetry to locate animals
- Censoring is a random event, and independent of mortality
- The relocation method (e.g. radio-telemetry transmitters) do not impact survival

The only tricky assumption is the statement about censoring, with regard to its independence of mortality. Unfortunately, it is usually difficult to validate this assumption, and mortality events could be predicted to cause radio failure at rates higher than mechanical failure or disappearance of the animal.

Censoring

To demonstrate censoring under Kaplan-Meier methods, we will consider an example of radio-marked northern bobwhite quail from Pollock et al. (1989). Eighteen individuals were radio-marked at Fort Bragg, North Carolina, USA in the spring of 1985. During the study, 6 individuals were found dead, and 5 disappeared (and were censored).

During weeks when no animals were censored, the estimation of s_t is straightforward. For example, in Week 3, 2 deaths were recorded while 18 birds were at risk. So, $s_3 = (18-2)/18 = 0.8889$.

Week (t)	No. at risk (t)	No. deaths (t)	No. censored	Newly added	s(t)	S(t)
1	18	0	0	0	1.0000	1.0000
2	18	0	0	0	1.0000	1.0000
3	18	2	0	0	0.8889	0.8889
4	16	0	0	0	1.0000	0.8889
5	16	0	0	0	1.0000	0.8889
6	16	1	0	0	0.9375	0.8333
7	15	0	0	0	1.0000	0.8333
8	15	1	1	0	0.9286	0.7738
9	13	1	2	0	0.9090	0.7035
10	10	1	1	0	0.8889	0.6253
11	8	0	0	0	1.0000	0.6253
12	8	0	1	0	1.0000	0.6253
13	7	0	0	0	1.0000	0.6253

From Pollock et al. (1989), with a correction in Week 9 made by Powell and Gale.

During weeks when censoring occurred, the number of animals at risk is modified by subtracting the animals that were at risk. For example, Week 9 started with 13 animals at risk, but 2 were censored. Even though we started the week with 13 at risk animals, our cohort must be modified (13 at risk - 2 censored = 11 at risk) to reflect the size of the cohort for which we know the final status. Therefore, we pretend the two censored birds were not there when Week 9 began. We note that one of the 11 at-risk birds died. So, $s_9 = (11-1)/11 = 0.9090$.

Figure 9.6: *Northern bobwhite in North Carolina, photo by Dick Daniels (public domain).*

During each week, it is possible to provide a 95% confidence interval by estimating the variance of the cumulative survival, S_t, at that period, t, for r_t animals at risk (Pollock et al. 1989):

$$\hat{\text{var}}(\hat{S}_t) = \frac{\hat{S}_t^2 (1 - \hat{S}_t)}{r_t}$$

And, we can provide a 95% confidence interval as:

$$\hat{S}_t \pm 1.96 \sqrt{\hat{\text{var}}(\hat{S}_t)}$$

If we create a figure of S_t through time, with the confidence intervals we see this pattern of survival for our cohort of 18 quail. We can see that the confidence intervals grow in size (less certainty in the estimate) as the sample size declines because of deaths and censoring (Figure 9.7).

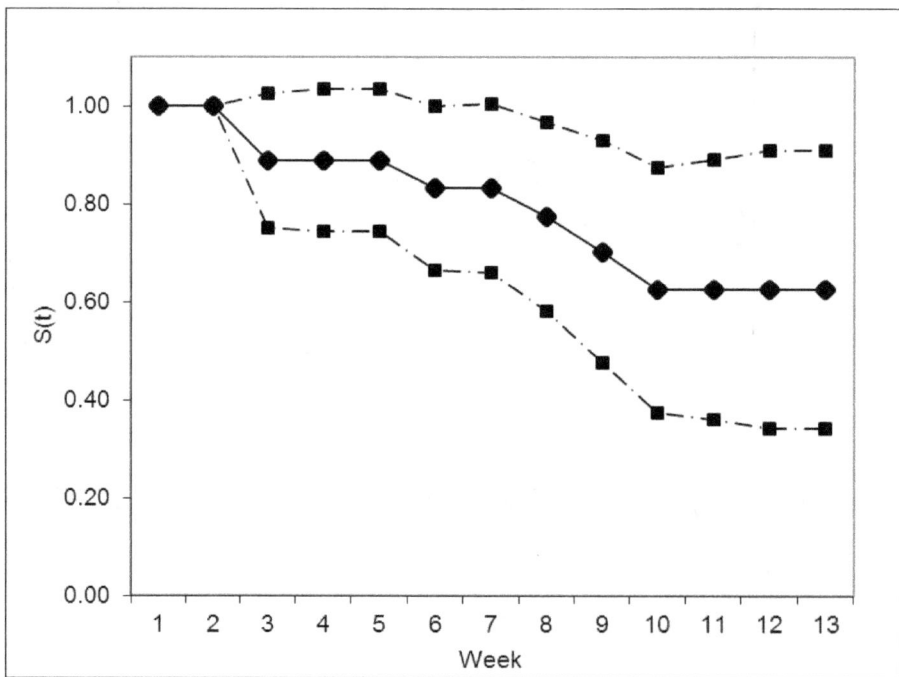

Figure 9.7: *Cumulative survival estimates, with 95% confidence intervals, from a Kaplan-Meier analysis of the Pollock et al. (1989) quail data.*

Kaplan-Meier methods use two types of censoring: **left censoring** and **right censoring**. To this point in the chapter, we've been using the term "censor" to indicate the most frequent type of censoring—right censoring. As the terms suggest, right-censoring means to "cut off" an animal's risk-presence on the "right side" or the latter time periods of the study. In contrast, left-censoring means to trim the risk-presence on the "left side" or the early time periods.

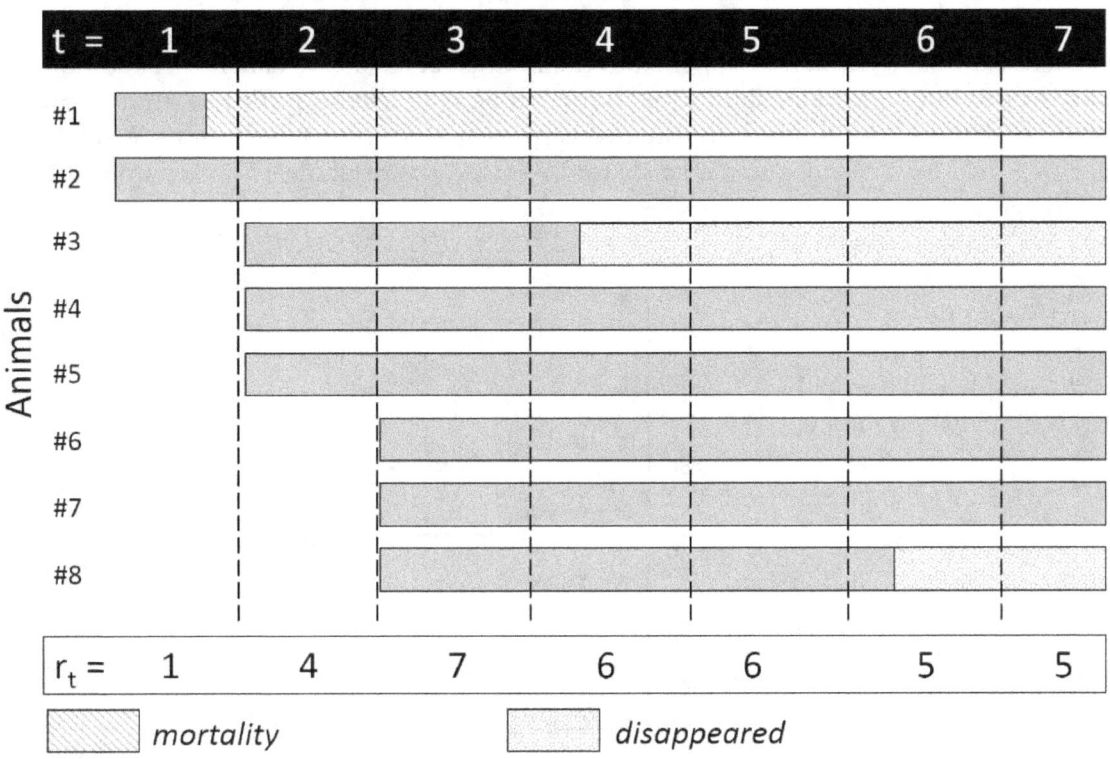

Figure 9.8: *Fates of a sample of 8 radio-tagged animals shown through 7 time periods, t. Mortalities and disappearances (to be right-censored) are indicated with stripes and light shading. The at-risk sample size, r, is shown at bottom for each time period.*

Figure 9.8 shows an example where both types of censoring are needed. To the right side of the figure, we see the Animal #3 and Animal #8 disappear from the sample. As we have done in the previous quail example, those individuals would be right-censored and would not be considered in the at-risk cohort.

But, we also see that we have a unique problem with Animal #1 and Animal #2—both are the only animals at risk during Week 1, and Animal #1 is found dead. Because of the small sample size, $s_1 = 1/2 = 0.500$. And, S_t is cumulative, so S_t will never get larger than 0.500. The Kaplan-Meier survival function (S_t) can only go down. Of course, we would guess that the animal's survival during one week is most likely not 0.500—so the proper thing to do is to left-censor the data. We eliminate Week 1 from the analysis and we start with Week 2.

Left-censoring is a consideration because we base our estimate of S_t and $\text{var}(S_t)$ on the sample size in the interval. In essence, we continually divide the data into smaller piles. Although we typically worry only about the effect of small sample size on the variance, we can now see that the estimate for S_t can also be affected.

We can describe the process of left-censoring as the act of determining "when to start the analysis". If biologists are worried about the effect of the capture event on survival, they may left-censor an individual for a few days before officially adding it into the 'at risk' pool—giving the animal time to recover from the capture event. **We recommend that you make decision rules about when you will left-censor before you begin your field work.**

Figure 9.9 shows what <u>could</u> have happened to the Pollock et al. (1989) data if two additional birds were tagged during a preliminary Week 0 (we stress: *this is purely hypothetical*—not a part of the Pollock et al. (1989) data). If one of the two early-tagged birds had died during Week 0, the survival rate would have plunged immediately to 0.500, and the resulting 13-week survival would be much lower (about 38%) than the 63% survival reported by Pollock et al. (1989; note: if we left-censored this hypothetical data set—to address the problem, we would see the results shown in Figure 9.7)

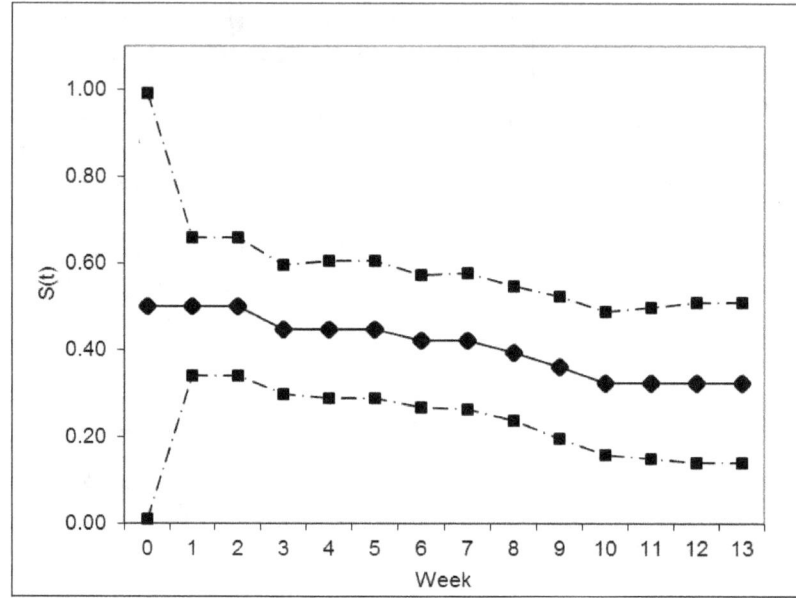

Figure 9.9: *Hypothetical result from a Kaplan-Meier analysis of the Pollock et al. (1989) data if 1 of 2 tagged individuals died during week 0. Left-censoring can be used to avoid this problem.*

Left-censoring, because of small samples during a staggered entry process, is only a concern when performing the time-specific analyses that are characteristic of Kaplan-Meier. Analyses with the Mayfield method or the general binomial model, described later in this chapter, have the option to simply pool the limited data from the early periods with other data, unless a time-specific model is required.

Capture histories for known fate data

The software you use for analysis will dictate the exact form required for your input files. But, it is worth describing the general form that is used to describe known-fate data for capture histories. The type of input that is a standard for program MARK is called **live-dead**, or LDLDLDLD. This type of capture history varies from the simple single-digit entries required for estimation of population size in Chapter 8 or CJS-type survival in Chapter 10.

Live-dead encounter histories use two digits for each occasion. The simplest capture history, for an animal surviving through all 5 time periods of a 5-occasion study is:

10 10 10 10 10

We place a "1" in the first column of the each occasion to indicate that the animal was alive and at risk in all 5 time periods. Thus, we know that this animal was tagged during (or before) the first time period. We place a "0" in the second column of each occasion to indicate that this animal never died. If the animal had died in the second time period, the capture history would read:

$$10\ 11\ 00\ 00\ 00$$

And, we see that the animal was at risk during the first two time periods, but was found dead during the second time period. By definition, the dead animal was not at risk during the last three time periods!

You try it!

Can you explain what happened in these capture histories? Describe the status of each individual. Answers are at the end of this chapter.

> Critter A: 00 00 00 10 10
> Critter B: 10 10 00 00 00
> Critter C: 00 10 10 11 00

Found it! A little more about right-censoring

Suppose one of the individuals in your study disappeared for more than one time period, and then you found it alive. Such a discovery is always a good feeling, when you're hot and tired in the field (or cold and tired, if you're tracking polar bears). The prodigal critter has returned, and all is well with the world again, right?!

So, it's time to construct your capture history and you are torn. The capture history that directly reflects the animal's participation in your study is 10 00 00 10 10, because you didn't know where it was during time 2 and 3. However, you found the animal alive—so logic tells you that it was alive while you could not find it. So, perhaps you should code the capture history as 10 10 10 10 10?

The former approach is correct. **Do not fill in the time periods that you did not record the animal "at risk."** Live animals are more likely to be re-encountered than dead animals, so by filling in 'survival' information after you find a live animal, you positively bias your survival estimate.

The general, binomial approach to known fate

Earlier in this chapter, we noted that the binomial estimator (with y successes in n trials) for known-fate survival was

$$\hat{S} = \frac{y}{n}$$

How can we generalize the idea of the Kaplan-Meier estimator, with survival estimated for specific intervals? What approaches do modern software use?

To build in multiple time intervals, we can make our estimator time-specific for time period i:

$$\hat{S}_i = \frac{y_i}{n_i}$$

To think of known-fate analyses in a likelihood framework, we must stipulate what can happen to a tagged animal before the next time period:
- The animal dies and is found dead.
- The animal survives and is found alive.
- The animal, known to be alive at the beginning of the time period, disappears and is censored.

It is important to note that the third possibility, censorship, is not actually a third "status"—it is simply a condition that requires the removal of the animal from the "at risk" sample for that time period. Because of this, we can still use the binomial approach shown above, with "alive" indicating "success" and "dead" indicating a "failure". That is, we do not need a third "censored" category and we do not estimate the probability of censorship.

For a given time period in a set of time periods, i, that have n_i animals alive and at risk during the time period, we can write the likelihood statement for known-fate survival (S_i) as:

$$L(S_i \mid n_i, n_{i+1}) = \binom{n_i}{n_{i+1}} S_i^{n_{i+1}} (1-S_i)^{(n_i - n_{i+1})}$$

Let's apply this to the Pollock et al. (1989) quail example from earlier in this chapter. An abbreviated table is shown here. We can see that the number of animals at risk, n_i, during each time period would be: $n_1 = 18$, $n_2 = 18$, $n_3 = 18$, and $n_4 = 16$.

Week (t)	No. at risk (t)	No. deaths (t)	No. censored	Newly added	s(t)	S(t)
1	18	0	0	0	1.0000	1.0000
2	18	0	0	0	1.0000	1.0000
3	18	2	0	0	0.8889	0.8889
4	16	0	0	0	1.0000	0.8889

From Pollock et al. (1989), first 4 weeks.

The likelihood used for S_3 is:

$$L(S_3 \mid n_3, n_4) = \binom{n_3}{n_4} S_3^{n_4} (1 - S_3)^{(n_3 - n_4)}$$

$$L(S_3 \mid n_3 = 18, n_4 = 16) = \binom{18}{16} S_3^{16} (1 - S_3)^{(2)}$$

Using the techniques we covered in Chapter 3, we could estimate S_3 from the above likelihood statement. Similarly, we could estimate all other S_i's.

Conclusion

The structure of known-fate models for survival are relatively simple, because we do not need to worry about encounter rates—all animals are assumed to be located with 100% success each time period. Mayfield uses the concept of exposure days (days a nest or some organism is exposed to potential predators or other source of mortality) typically to assess nest survival, while Kaplan-Meier uses a similar concept of looking at the number of animals "at risk" during a particular time period to examine variation in survival through time. If animals are not located, they are right-censored from the sample. Mayfield and Kaplan-Meier methods provide simple and historic examples of the applications of known fate methods. The generalized, binomial approach to known-fate data allows a more flexible assessment of variation in survival because it provides the avenue to investigate the effects of covariates on survival using linear models.

Figure 9.10: *A radio-marked leopard (Panthera pardus) surveys its territory in Mashatu Game Reserve in eastern Botswana; photo by Andrei Snyman, used with permission.*

Answers: You try it!

Can you explain what happened in these capture histories?

Critter A: 00 00 00 10 10
Critter B: 10 10 00 00 00
Critter C: 00 10 10 11 00

Answers: "Critter" A was not tagged until time period 4, and then it survived through the end of the study. Critter B was tagged in time period 1, and survived through time period 2. However, it is not listed "at risk" in time period 3, which suggests that Critter B disappeared during time period 3 and was right-censored. *We note that the survival information provided by Critter B during times*

1 and 2 is still used in the analysis. Critter C was not tagged until time period 2, and it survived until time period 4 when it was found dead.

References

Fries, J. F. 2002. Aging, natural death, and the compression of morbidity. Bulletin of the World Health Organization 80: 245-250.

Kaplan, E. L. and P. Meier. 1958. Nonparametric estimation from incomplete observations. Journal of the American Statistical Association 53: 457-481.

Lang, J. D., L. A. Powell, D. G. Krementz, and M. J. Conroy. 2002. Wood Thrush movements and habitat use: effects of forest management for Red-cockaded Woodpeckers. The Auk 119: 109-124.

Manolis, J. C., D. E. Andersen, and F. J. Cuthbert. 2000. Uncertain nest fates in songbird studies and variation in Mayfield estimation. Auk 117: 615-626.

Mayfield, H. 1961. Nesting success calculated from exposure. Wilson Bulletin 73: 255-261.

Pollock, K. H., S. R. Winterstein, C. M. Bunck, C. M., and P. D. Curtis. 1989. Survival analysis in telemetry studies: the staggered entry design. The Journal of Wildlife Management 53: 7-15.

Skalski, J. R., K. E. Ryding, and J. Millspaugh. 2005. Wildlife demography: analysis of sex, age, and count data. Academic Press.

For more information on topics in this chapter

Conroy, M. J., and J. P. Carroll. 2009. Quantitative Conservation of Vertebrates. Wiley-Blackwell: Sussex, UK.

Cooch, E., and G. White. 2014. Chapter 16: Known-fate models. *In* Program MARK: a gentle introduction, 12th edition, Cooch, E. and G. White, eds. Online: http://www.phidot.org/software/mark/docs/book/pdf/chap16.pdf

Rotella, J. 2014. Chapter 17: Nest survival models. *In* Program MARK: a gentle introduction, 12th edition, Cooch, E. and G. White, eds. Online: http://www.phidot.org/software/mark/docs/book/pdf/chap17.pdf

Williams, B. K., J. D. Nichols, and M. J. Conroy. 2002. Analysis and management of animal populations. Academic Press, San Diego.

Citing this primer

Powell, L. A., and G. A. Gale. 2015. Estimation of Parameters for Animal Populations: a primer for the rest of us. Caught Napping Publications: Lincoln, NE.

Ch. 10

Cormack-Jolly-Seber: estimating apparent survival

"...Nature, red in tooth and claw"
-- Alfred, Lord Tennyson

"I've always believed in survival."
-- Hugh Leonard

Questions to ponder:
- *What is the difference between survival and apparent survival?*
- *How can survival be estimated from the information found in a set of capture histories?*
- *Is it possible to use a 2-occasion capture exercise to estimate survival?*

A basic framework for survival of marked animals

In the previous chapter, we covered the use of mark-recapture in a closed population form to estimate N, or population size. Now, we'll extend the general ideas of sampling without certainty of catching the animal again ($p<1.0$) to **open populations**—populations for which the time period of consideration allows births and deaths to occur. The type of data structure on which we focus here is the basic type of survival analyses that biologists use to assess survival of marked individuals which are not radio-marked—that is, individuals with leg bands, tags, or other marks. In contrast to Chapter 9, animals with these types of marks have **unknown fate**, as we cannot follow them and know their fate unless we capture them again.

We call this method "CJS" after Richard Cormack, George Jolly, and George Seber—three people who developed the method separately in 1964 (Cormack) and 1965 (Jolly and Seber).

The typical mark-recapture study for CJS-type analyses consists of a population identified for study and a method employed to capture animals. Almost any method can be used: nets, traps, pit falls, dart guns, or whatever can be used to capture a representative sample of the animal species of interest.

Those animals then receive some kind of mark that allows individual identification. Although population size can be estimated using marks that are not individually unique—such as paint splatters on iguanas or a fin clip of a fish—**CJS apparent survival estimation requires unique marks on individuals**. Visual sightings can be used for re-captures of animals if the proper kind of visual markers is used.

The mark-recapture project must be extended for at least 3 time periods. Later, we will discuss why you can't stop with two sampling occasions if you have interest in survival. During each time period, animals are captured, identified (if already marked) or newly marked, and then released back into the population, alive.

Time periods need to be long enough to make sense for the species of interest---some mortality needs the opportunity to occur during that time period. So, for most larger fish and wildlife, biologists might use month or three-month or six-month or annual intervals, as examples.

Once an animal is marked and released, three potential things can happen to it if we look forward to the next time period (Figure 10.1):

- First, an animal can survive through time period 1 and be captured at the start of time period two. That animal, has the capture history of "11".
- Second, an animal that survives through time period 1 may be alive during time period 2, but not captured in that time period. So, it would have a capture history of "10".
- And last, an animal may be released at the beginning of time period one and it may die of various causes during the first time period. It is not available for capture. So, it would also end up with a "10" capture history.

A key point is that, with CJS-type data, we cannot distinguish between the last two possibilities. That is, we have no way of knowing if the third animal died or not. It is not wearing a radio-tag. And, we did not find it dead. So, animals not captured again simply disappear. And, this is one reason that we need at least three time periods to estimate survival. If we stop at two time periods, we are left with an unsolvable dilemma—and no way to separate survival and capture rate for these animals with "10" histories (Figure 10.1).

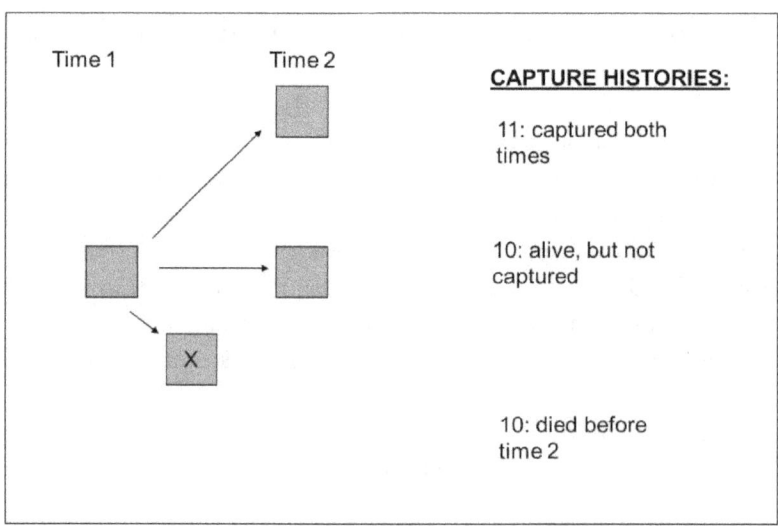

Figure 10.1: *Capture histories for marked animals with three possible transitions to capture and survival status in Time 2 after release in Time 1.*

We should also note at this point that the survival we estimate from CJS-type data analyses is labeled **'apparent survival.'**

A fourth alternative is actually possible here…that is, the animal could have remained alive, but left the study area. That would also result in a "10" capture history. There is definitely not a way to separate out the movement part of this from the mortality part when using the CJS framework. So, we assume that emigrated animals are dead, and we use the label 'apparent' survival in our publications to indicate that it is actually a conservative estimate of survival. That is, the true survival rate may actually be higher than the apparent survival estimate we obtain from CJS if movement rates out of our study site are high. To our CJS model, a situation in which animals are moving away would cause the capture histories to appear as if many animals are dying, and survival will be estimated to be low. The best we can do is to call it 'apparent' survival. *Note: Many editors will insist on this label in journals to distinguish CJS-estimated survival from known-fate survival or other estimates of survival.*

$$1 \xrightarrow{\phi_1} 2 \xrightarrow{\phi_2} 3 \xrightarrow{\phi_3} 4 \xrightarrow{\phi_4} 5 \xrightarrow{\phi_5} 6 \xrightarrow{\phi_6} 7$$
$$\quad\quad p_2 \quad\quad p_3 \quad\quad p_4 \quad\quad p_5 \quad\quad p_6 \quad\quad p_7$$

Event timing in CJS models

It is worth a reminder of the timing of events that are assumed in the structure of a survival analysis. Animals are released IMMEDIATELY at the beginning of a time period. In fact, their capture and release are assumed to happen instantaneously and together the capture and release define the beginning of a time period.

So, the survival estimate, ϕ_1 ("phi 1"), here corresponds to the probability that an animal released at the beginning of time period 1 survives to the second time period. The capture probabilities are defined as the probability of being captured in a given time period, given that the animal is alive and available for capture. At the moment, this looks very sensible. It will become important to remember as our model structures grow more complex—for example as we head toward robust-design or multi-state models in future chapters.

NOTE: we will use the convention of other authors to distinguish between "true" survival, S, and apparent survival, ϕ. This distinction will become especially important in Chapter 13 for the robust design models.

Another thing to note, as we look at this CJS example, is that if you have 7 time periods, you will only obtain 6 survival rate estimates---and, as we will discuss momentarily---the last one is actually not estimable as a lone parameter. So, in effect, you will only obtain 5 survival rate estimates. This is important in study design and long-term planning. A student who wants to do a 3-year field project will actually only be able to estimate 2 annual survival rates, and the second is not completely estimable as a time-specific rate. We will come back to this concept later.

Cohorts working together

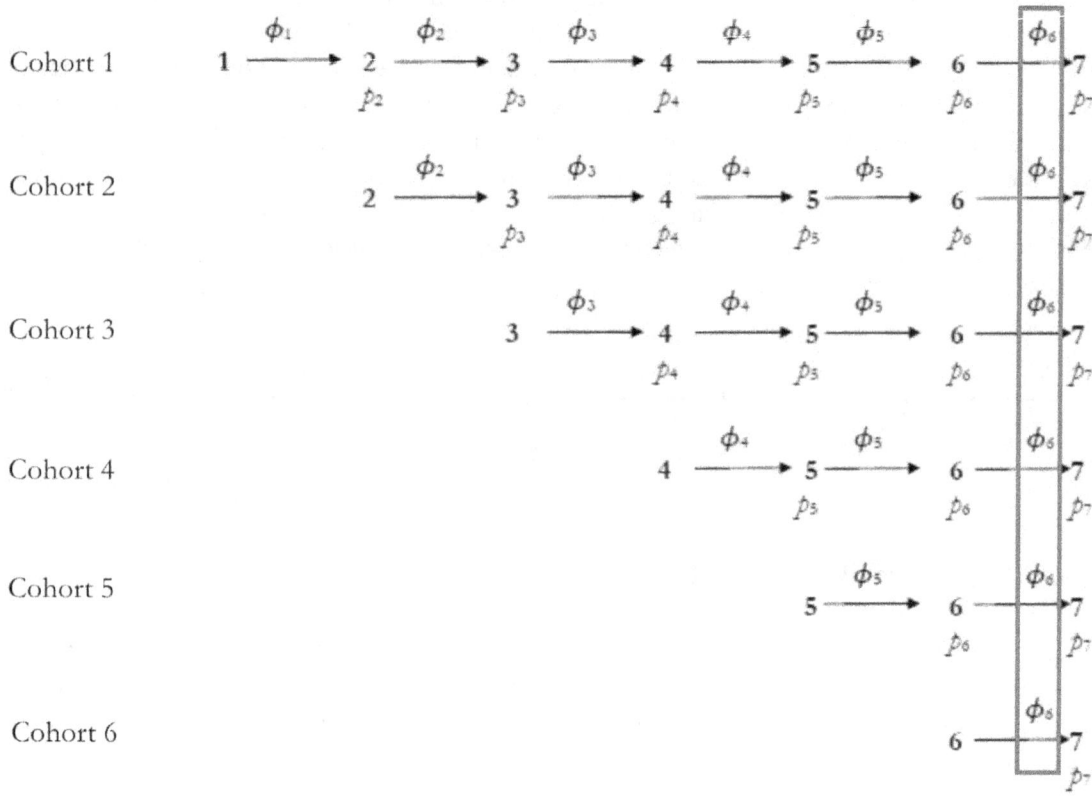

Figure 10.2: *Example of 6 cohorts of released animals, captured and recaptured during a 7-period study to estimate apparent survival (ϕ). Apparent survival between periods and capture during periods (p) are shown. Vertical box during period 6 indicates the logic that all 6 cohorts provide information for an estimate of apparent survival for the population during that time period, but survival is not estimable as a time-specific (e.g., ϕ_6) parameter for the last interval.*

We've previously discussed the notion of a cohort. In a multi-year study, cohorts will eventually exist in the presence of other cohorts that are released in subsequent time periods, and in Figure 10.2 we see how a series of cohorts work together in a study that continues to capture new animals through time. In time period one, the first cohort is released. In time period two, a second cohort of animals is released. The same thing happens during all time periods.

Notice that all animals—regardless of which cohort they are in—are used to estimate time-specific parameters. For example ϕ_6 (or survival through the sixth time period) can be gotten by using data from all six cohorts.

Probability statements

One of our goals is to become more comfortable thinking about events in mark-recapture data as probability statements. Here, we have a table of cohorts that are released and animals captured again, for the first time, in subsequent time periods.

Cohort released	Probability of recapture, first, in time period x		
	2	3	4
R_1	$\phi_1 p_2$	$\phi_1(1-p_2)*\phi_2 p_3$	$\phi_1(1-p_2)*\phi_2(1-p_3)*\phi_3 p_4$
R_2		$\phi_2 p_3$	$\phi_2(1-p_3)*\phi_3 p_4$
R_3			$\phi_3 p_4$

Let us start with a look at a released cohort of animals at time one: R_1. It is possible that these animals will be next captured in time 2 (a capture history of "11" when they are recaptured for the first time). Or, they may skip capture in time 2 and be next captured in time 3 (capture history of "101" at first recapture). Or, an animal may go uncaptured through time 2 and 3 and be next captured in time 4 (capture history of "1001"). So, for R_1, we can write out several probability statements that go with these three possibilities.

It may help to walk through the logic behind the probabilities. First, what **binomial events** must happen to an animal released in time 1 and next captured in time 2 (Figure 10.1)? We know that it is has to remain alive to be captured again. So, it has to survive through time period 1 to be captured in time 2. And, it has to be captured in time two. Those two events (survival and capture) BOTH have to happen, so we multiply the probabilities: $\phi_1 * p_2$.

Similarly, if an animal is released in time 1 and captured next in time period 3, we can write the probability of that event. The animal has to survive through time 1, then it must NOT be caught in time period 2, and then it must survive through time period 2, and then (finally) be captured in time period 3. So, you should now envision a string of 4 probabilities for that 'event.' We used $1-p_2$ to indicate that the animal is not captured in time 2. You can see a similar probability statement for time period 4. Work through it in your mind—it should make sense in terms of the order of binomial events that have to happen to be released in 1 and only seen again in time period 4: survive, not captured, survive, not captured, survive, and captured. We write that formally as: $\phi_1(1-p_2)*\phi_2(1-p_3)*\phi_3 p_4$

These probability statements are at the core of what the CJS models 'know' or assume about our system as we attempt to estimate these parameters from data that we collect.

CJS model assumptions

As with all data structures and models, CJS-type survival analyses come with some **assumptions**.

- Every animal has the same chance of capture, p
- Every animal has same probability of surviving (ϕ; phi) to the next sampling period
- Marks are not lost/overlooked
- Samples are instantaneous (short periods) and animals released immediately
- All emigration from the sample area is permanent (101 must indicate "1-p")
- Fates of animals (with regard to capture and survival) are independent of other animals

The first assumption simply means that we assume that our capture probability estimate applies to all individuals in the population—or at least within a group that we designate in our modeling with a specific estimate for p. For example, we could break the population into males and females, which could be allowed to have different estimates of p—but we assume then that all males have the same chance of capture.

We would call this a 'definitional' assumption, then. The assumption is because of the manner in which we estimate the p's. We acknowledge that we have no way to account for heterogeneity in capture probabilities within a population in the basic models. So, if we THINK there are differences, we need to sample within those groups with an appropriate sample size for each group (such as the males and females) to account for differences that may exist.

If we ignored this assumption and had disproportionate sampling (e.g., ¾ of sample were males and ¼ females)…then our estimate of 'p' for the entire population would be biased by the males and might affect the estimates of survival of the animals.

The assumption about instantaneous sampling and release bears some comment. We assume the sampling and release is instantaneous, and this seems obvious. But, what if you were doing annual survival of fish in the Missouri River and your sampling season was three months long (June, July, and August)? Some fish would be released at the end of those summer months and only have to survive 9 months until the next sampling season while some fish would be released at the beginning of the summer and have to survive 12 months until the next sampling season. In this Missouri River fisheries example, we are probably not going to have many problems by simply acknowledging the assumption in our analyses. But, what if your sampling period each year was 6 months long? Now, fish sampled in the first month have to survive 5 months longer to be available for sampling during the next year than fish sampled in the last month—and all are represented in the same manner in your capture history with a "1" for that year. You can see the assumption of instantaneous sampling is being stretched, in this example. Or, to push this point even further, what if sampling occurred during the entire year!? Collapsing all of your data into one, annual interval is certainly going to be problematic because of the instantaneous sampling assumption. One approach to avoid this problem would be to sample during two or more, 3-month time periods instead of 6-month or 12-month sampling periods. Generally, the sampling periods need to be much shorter than the time periods used in your survival model.

A note about the emigration assumption: we have already described that the CJS structure has no way to distinguish an animal that leaves the study area with one that dies. And, that is because of the design of our study—CJS-type data comes from a single study area. We aren't capturing animals in other places. So, all animals are 'dead' to the model when they emigrate. Because of that, animals that emigrate and then immigrate back to the area in a subsequent time period become problematic. A "101" capture history is assumed to mean that the animal was *on site* during time 2, but not captured. A lot of movement in and out starts to affect capture probabilities, and may mess up the estimates for this model structure. If you have a lot of movement, you might like to explore the robust-design or multi-state models in future chapters—to document and take advantage of movement information (which also means revising your study design to have multiple sites for multi-state data).

CJS models are relatively simple—they assume movement is not occurring. Remember that p is defined as the probability of capture, given that an animal is on-site and available for capture. You are breaking this assumption if your study animal has moved to Canada and your study site is in Mexico.

Last, the model structure treats all individuals as separate, independent creatures. We know that nature is messy and animals clump together and experience the same mortality risks as others, from time to time. For example, breeding pairs of songbirds or male deer in large groups during the non-breeding season, or fish in a school. Again, this assumption becomes a problem if *too much* pooling occurs. So, we need to be careful how we sample. Perhaps we mark only one individual from a given group if we are worried about the effects of the group on the individual.

These are all things that are important to consider as you design a study.

Logic behind the CJS model structure

How does the CJS model estimate survival from your data?

We can start with the closed-population logic seen in the Lincoln-Petersen estimator for closed populations (Chapter 8). In that 2-sample occasion, Lincoln and Petersen noted that there are two ratios that *should* be the same---the portion of the population that you capture during the first time period, n_1/N, and the portion of the second sample that is marked, m_2/n_2.

$$n_1/N = m_2/n_2$$

That is, all animals in the population have the same probability of capture, so the marked portion of the second sample should be equivalent to the portion that was sampled in the first sample. CJS simply extends that to an open situation, with the same logic: all individuals (whether previously marked or not) have the same probability of capture.

$$M_i/N_i = m_i/n_i$$

We introduce M_i to indicate (as with M_t in the Schnable estimator in Chapter 8) the number of animals that have been marked *previously* to the start of any capture period. And, m_i is defined as the number of marked animals captured in sample i, and n_i is the number of animals (marked + unmarked) in sample i.

So, Cormack, Jolly, and Seber postulated that the portion of the population that has been marked previously (M_i/N_i) should be equal to the portion of marked animals in the current sample (m_i/n_i).

The next step in this, and a fairly simple, straightforward one....is to note that survival, ϕ, is simply the portion of marked animals at the current time that are alive in the next time period.

$$\phi = M_{i+1} / M_i$$

So, we've marked 50 animals by time period 6 for example. By time period 7, 40 of them are alive. So, survival is simply 40/50 or 80%.

Ignoring <u>new</u> marks, the only way that M_i can change is to have animals die or leave the study area. Furthermore, we have no means of distinguishing emigrants from deaths, so they both are considered deaths.

This seems pretty logical. But, of course, the problem is that these animals are not radio-marked or tethered to our research station with a cable. We normally *do not* know when animals leave or die. We just know when we mark them and when we catch them. So, the simple logic gets a bit more complex, because **we have to use things we do know to estimate what we don't know**.

In fact, as Cormack, Jolly, and Seber postulated, we actually do know quite a bit about a sample of marked animals…if we just look for evidence 'hidden' in their capture histories.

So, we have a group of 'sufficient statistics' as they are termed…essentially, some parameters that we need to estimate or calculate to estimate the parameter in which we have interest (i.e., survival).

Here is the list of these **sufficient statistics**. We can think about a set of captured animals with their capture histories as we review these:

- R_i : the number of animals released at time i. This is our **cohort**, and it becomes a critical group of animals.
- r_i : the number of animals in the cohort from time i that are captured again…at any point in the future. So, if we release animals in time period 4, r_4 would be animals released in time 4 that are captured again in time 5, 6, 7, 8, 9, 10, etc. until the end of the study.
- z_i : animals captured before i, not captured in i, but captured later after i. Admittedly, this is the oddest sufficient statistic, and one that may take some time to grasp. It is the

number of animals captured before the current time period which were *not* captured in the current time period *but were captured* in a later time period. This may be easier to visualize when we look at some capture histories below.

- M_i-m_i : the number of marked animals (M_i) that *are not* sampled in the current time period (remember, m_i is the number of marked animals that *are* captured in the current time period).

We can relate these sufficient statistics in an equation:

$$r_i/R_i = z_i/(M_i-m_i)$$

This equation is an equality of two ratios, and we will explore this momentarily. But, for now, just recognize that we have an equation that has M_i in it (which we need to estimate survival across periods). And, all other parameters in that equation are *things that we can find in our data*—by looking at our capture histories. So, we are now able to estimate M_i for the current time period! This is the brilliance of Cormack, Jolly, and Seber. We can estimate M_i; and if we can do that for two time periods in a row, we can estimate survival!

Sufficient statistics: an example

To help us look at the basic components of an apparent survival analysis-- R, r and z—let us examine a set of 9 capture histories. As a reminder, each row is a capture history for one individual over 10 capture occasions. A "1" indicates the individual was caught in time i, and a "0" indicates the individual was not captured in time i.

Let's look especially at time period 3, highlighted here by the red vertical box. We can see which of the 9 individuals were captured in that time period---the 2nd and 3rd individual, as well as the last two individuals in the set. No other individuals were captured during this period.

So, we can calculate $R_3 = 4$. We captured and released (alive) 4 individuals. That's the easiest statistic to calculate.

The value r_3 is equal to the number of R_3 that are captured again later. Of these 4 'critters' released in time 3, three are seen again—we can easily look to see if there is another "1" in the capture history after time 3. And, we can also see that the second individual in this set released in time 3 was never observed again. Therefore, $r_3 = 3$.

But, z_i is our 'trouble-maker'. This is the statistic that makes our head 'swim'. To review, z_3 is the number of animals *not caught in time 3* but *captured before and after time 3*. We start by looking for 0's in the capture histories during time 3—to indicate animals not caught in time 3.

```
1001011000
1010000000
1011100001
1000000000
1000001000
0100000100
1001001001
1011001100
1111111111
```

Figure 10.3: *A set of 9 capture histories used to estimate apparent survival. The capture status (0 or 1, not captured or captured) is emphasized for time period 3.*

There are 5 of them: the 1st, 4th, 5th, 6th, and 7th individuals in the set of 9. Now, we look at each one of those. Were they captured before time 3? Well, it turns out that…yes…they were all captured before (we can see a "1" in their history previously). And, finally, were they captured after time 3? Only individual 4 was not captured later. But, all the others were captured. So, z_3 = 4.

We note that z_i is a complicated 'little' statistic, but it turns out to be important for the logic of the CJS estimator. Make sure you understand z_i!

The grand equality

Let's return to our equation from earlier—the critical step in the CJS estimation process, as we try to estimate M_i:

$$r_i/R_i = z_i/(M_i - m_i)$$

If you are a visual learner, the flow chart in Figure 10.4 may help you appreciate the logic in this equation. Why are these two ratios set aside as equal and what is the logic?

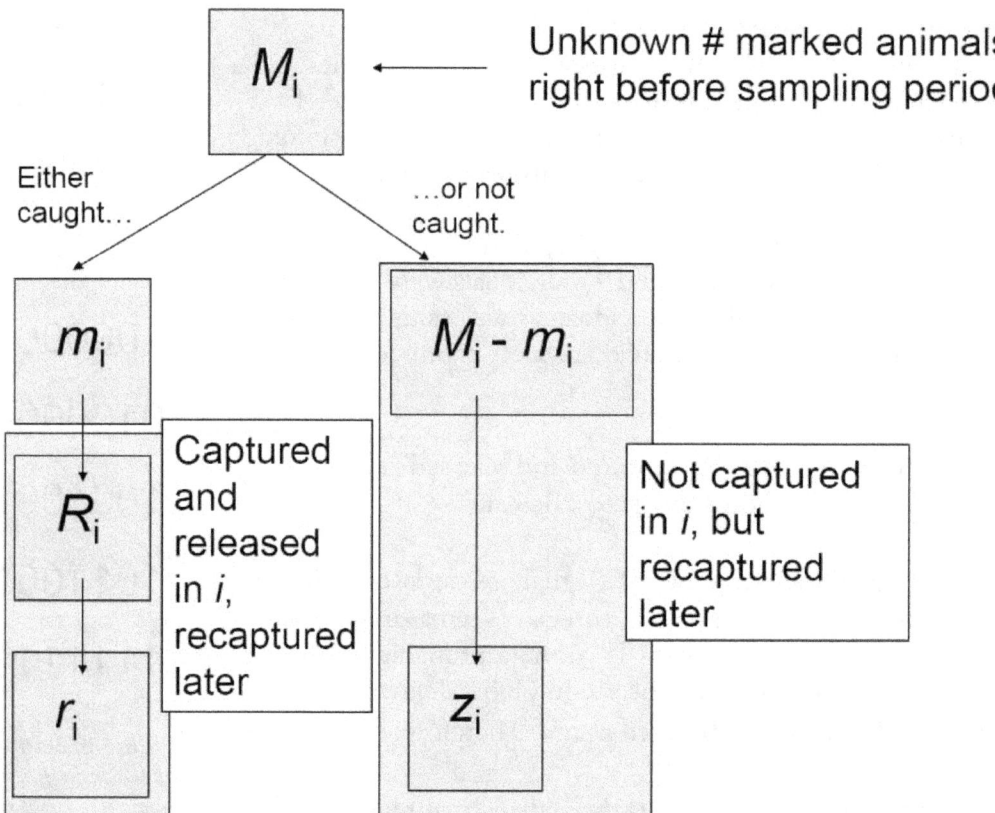

Figure 10.4: *Depiction of the 'grand equality' that allows estimation of apparent survival from CJS-type data. The unknown number of marked animals alive before a sampling period (M_i) will either be caught (m_i) or not caught ($M_i - m_i$) in the sampling period, i. The probability of capture in subsequent time periods is the same for both groups of marked animals, regardless of capture in time period i.*

As we evaluated Figure 10.4, consider M_i, the number of marked animals you've marked in the population to this point that are still alive. For our discussion, let's say we say we are in time period 6.

So, consider M_6. You know how many animals you have marked, but you don't know how many have died (and thus you don't know M_6, the number of marked animals alive before time period 6).

We only know how many we've marked.

So, of those unknown number of marked animals that are still alive out there, there are only two things (a binomial event) that can happen to them during each time period! They can either be caught or they can be 'not caught'.

For time period 6, then let's divide the 'pile' of unknown marked animals into two 'heaps of critters'—those caught (these will go on the left) and those not caught (these go on the right) during time period 6 (Figure 10.4). We can now start to travel down the two forks of the diagram, and if you are especially alert you will notice that the two sides of the diagram also contain statistics that match the two sides of our equation, above!

Let's follow the side that we know something about—because we caught them during time period 6. These are designated as m_6. We are going to release R_6 of those in time 6 and we are going to recapture a certain number of them (r_6) after time period 6. Thus, the left side of our equation is that ratio: r_6/R_6. Essentially, this is the long-term capture probability of the released animals from time period 6 until the end of the study.

Now, let's go to the right side of the diagram and the right side of our equation, above. These are the marked animals that we did not catch during time 6. Again, we don't know how many we didn't catch, because we assume some of our marked animals have died without evidence during the past few time periods. But, we can enumerate these dead animals (unknown in number) as $M_6 - m_6$. We don't know what that number is, but we can stipulate it, theoretically and (importantly), we do know m_6.

Now, we see our 'friend', z_6. z_6 is a statistic that tells us something about the animals not captured in the current time period (time 6 in our example). We know that there are some animals out there not captured, because we can look at the capture histories and see that there were animals out there that were captured before time 6. And, for z of them we know that they were also alive after time period 6...because we can see they were captured later. By definition, if you were an animal captured before time 6 and after time 6, but not during time 6, I know you were alive in time 6...even though I did not record your capture. By logic and the assumptions of the CJS model structure, you had to be alive in time 6 if you were captured before and after time 6!

So, that number of 'known alive' critters not captured during time 6 can be thought of as a proportion of marked animals that are alive in time 6 but not caught.

Furthermore, that proportion should be equal to the portion of animals *known to be alive* in time 6 and recaptured later….because they have the same chance of being captured again. Is your head 'swimming' yet? It's relatively simple when you work through it, but it takes a bit of time to get your head around this…We warned you!

> **The key:** *given the animal is alive, it is assumed to have the same probability of being captured in a future time period, regardless of whether it was captured in time i.*

Now that we can manually find the statistics we need from our capture histories, we can be grateful that our software packages will do this for us in the future! And, we can rearrange the equality to estimate the parameter of interest, M_i:

$$M_i = m_i + (R_i * z_i)/r_i$$

For another reminder, we get survival estimates by using the estimates of M in subsequent time periods:

$$\phi = M_{i+1} / M_i$$

The logic of the formula shown here is that the number of alive, marked animals in the population (M_i) at a point in time is equal to the number of marked animals that we capture in the given time period (m_i), plus the number of marked animals that we don't catch, but who are still alive and still in our study area: $(R_i * z_i)/r_i$.

We really can get a lot of information from looking at capture histories, can't we? You didn't think it was magic, did you?!

Maximum "Eye-klihood" Estimation?

In a previous chapter, we discussed maximum likelihood estimation, or MLE. The general idea of MLE is that the process finds the most likely estimate for a parameter, given your data. Although at times this sounds like wizardry, our goal here is to cut through the haze and provide a glimpse of the logic behind the formulae. So, let's try a simple logic exercise—use your eyes to suggest some general answers.

Suppose a biologist conducts a seven-year, CJS-type capture-recapture study of three different populations of animals (they don't even have to be the same species). A portion of the capture histories (*we will assume the portion that is shown is representative of the entire captured sample*) are shown in Figure 10.5.

Study #1	Study #2	Study #3
10100000	11100000	10100101
10010000	11110000	10010101
01010001	01110101	01010101
01001000	01011000	01001001
01100000	01110000	01100010
01000000	01100000	01000001
10000000	11000000	10000010
00100100	00110100	00100101
00101100	00111100	00101101
00010100	00011100	00010101
00011000	00011100	00011010

Figure 10.5: *Three sets of capture histories from different hypothetical studies to provide visual appraisal of relative survival and capture probabilities between studies.*

Look at the capture histories closely. There are some differences that should start to appear to your eye. Look especially for evidence that your data provide for the length of time that animals remain alive, and also look for evidence of capture probability.

Start by comparing Study #1 with Study #2. What do you see? With regards to survival, do animals in one population appear to live longer than the other? No. Animals in study #1 are typically in the data set for about 3-4 years before dropping out (although the third individual in the set lived for at least 7 years, correct?). For animals in study #2 they appear to be roughly the same as #1. What is different between study #1 and study #2? You should see that there are many more instances of "11", "111", or "1111" in Study #2—that is, capture probability seems to be higher in Study #2, because animals are routinely captured many times in sequence after their initial capture. Very simply, there are fewer "0" entries in study #2, which means that the biologist had a relatively higher probability of capturing her study animals.

To our eye, we can see that, most likely (to use the MLE phrasing), capture probability is higher for Study #2. And, although our eye cannot tell us whether $p = 0.36$ or $p = 0.48$ or $p = 0.88$, we do know that $p_{study\#1} < p_{study\#2}$.

Now, let's compare Study #1 with Study #3. What do your eyes see here?
Is capture probability different? No, p doesn't appear to vary (at least noticeably) between Study #1 and Study #3. But, what about survival? Would you expect survival to be different if you conducted an analysis of the two data sets?

Yes, survival certainly seems to be much higher in Study #3. In Study #3, we see evidence (based on our capture histories) that animals typically survive 5, 6, or 7 years. And, as we noted before, the capture histories in Study #1 show evidence for animals surviving 3 or 4 years after initial capture (with that one exception of 7 years).

So, to our eye, we can postulate that survival is "most likely" higher for the species of animal that was captured in Study #3, relative to the species captured in Study #1.

We'll call this *Maximum Eye-klihood Estimation*—using our eye to look at our data and using our brain to make relative comparisons. "MEE" is a technique at which you should become adept. You should always *look* at your capture histories *before* submitting them to some package of software for analysis!

Estimable parameters

It is necessary to pause and discuss situations for which you cannot estimate a parameter (or may have trouble doing so).

First, if your data are sparse (few captures) and if you are using a time-specific model (asking for time specific estimates of survival), you may see bogus estimates of parameters or estimates for their variance in your analysis' output. That is, if you have 0's for many of the sufficient statistics for a given time period (r_i, z_i, or m_i), that will cause problems in the time-specific estimates. What can you do? If you believe a time-specific model is biologically or otherwise important, then try to estimate the time-specific parameters. But, be ready to collapse survival to a 'constant' parameter (that is, not time-specific) if you see problems with the estimates or variance for estimates in the time-specific models.

Note that this problem has implications for sampling as well. We need to have sufficient effort to get enough captures and recaptures of marked individuals during each time period.

Secondly, because of the structure of the models, it is *never* possible to estimate the last time period's survival probability in a CJS-type model structure. It turns out that you can only estimate the product of the capture rate and survival, which we loosely label as a 'return rate'— or the probability of seeing an animal again (surviving and being captured).

This may or may not be a problem to you. But, this problem does affect short-term, time-specific analyses that are often used by MS or PhD students who really want time-specific survival estimates.

To showcase why it's not possible to estimate the last period's survival rate, let's take the example of a two-time-period study. Now, we already said earlier that we need three time periods for a CJS study…to be able to get a "z" statistic. So this example will 'poke a nail in the coffin', so to speak.

> *Let's assume you are a MS student, and you have two field seasons to 'make your mark' on the world by estimating annual survival for a rare species. You work very hard to release 100 animals in time period one and you work equally hard to capture 60 of those marked animals during the second time period.*

Here's the problem encountered by any software package that employs a maximum likelihood estimator to get your survival estimates. There are literally an infinite number of combinations of values for ϕ and p that will result in 60 animals being sampled:

- You could have 100% survival and 60% capture probability.
- Or, you could have 80% survival and 75% capture probability.
- Or, you could have 60% survival and 100% capture probability.
- Etc.

The MLE is going to literally blow up. Well, perhaps not "literally", but MLE cannot work in this situation. There is no 'most likely value given the data'.

The solution: you need more data from additional time periods to estimate ϕ and p separately. Or, you can change to a model with constant recapture rate to obtain the last survival rate (now, p is the same across all time periods, and the other time periods are providing the information needed to estimate p—leaving the model with enough information to estimate ϕ for the last period).

However, in a model with time-specific survival and recapture probabilities, we get an estimate that the product of the two (ϕ and p) is 60%. That might provide useful information, to some extent, but it's not the apparent survival estimate that we wanted. This is important to consider when you design a study! Essentially, you always need one additional year of capture data to estimate annual survival and recapture rate(s).

Conclusion

Open populations are often more biologically relevant than closed populations because we are often interested in the probability of survival over time. The time periods used in CJS-type models however, need to fit the life history of the species of interest—longer periods for species with longer life spans and shorter periods for species with shorter life spans. CJS-models assume that capture probability is not 100%, so the models include parameters for capture probability and survival probability during each time period. A set of capture histories is used as the input data for CJS-type analyses, and we can calculate several 'sufficient statistics' from summaries of the capture histories that allow eventual estimation of survival.

CJS-type models estimate 'apparent survival' because the models do not include any movement parameters; thus, an animal that emigrates is considered to have died. Apparent survival estimates are considered conservative estimates of survival, as some animals (assumed to have died) are probably still alive but outside of the study area. If there appears to be a lot of movement, you may wish to consider the use of either the robust-design or multi-state models (Chapters 12 and 13) to document and take advantage of movement information.

For more information on topics in this chapter

Amstrup, S. C., T. L. McDonald, and B. F. J. Manly. 2005. Handbook of capture-recapture analysis. Princeton Univ. Press: Princeton, NJ.

Conroy, M. J., and J. P. Carroll. 2009. Quantitative Conservation of Vertebrates. Wiley-Blackwell: Sussex, UK.

Schwarz, C. J., and A. N. Arnason. 2014. Chapter 12: Jolly-Seber models in MARK. *In* Program MARK: a gentle introduction, 12th edition, Cooch, E. and G. White, eds. Online: *http://www.phidot.org/software/mark/docs/book/pdf/chap12.pdf*

Williams, B. K., J. D. Nichols, and M. J. Conroy. 2002. Analysis and management of animal populations. Academic Press: San Diego.

Citing this primer

Powell, L. A., and G. A. Gale. 2015. Estimation of Parameters for Animal Populations: a primer for the rest of us. Caught Napping Publications: Lincoln, NE.

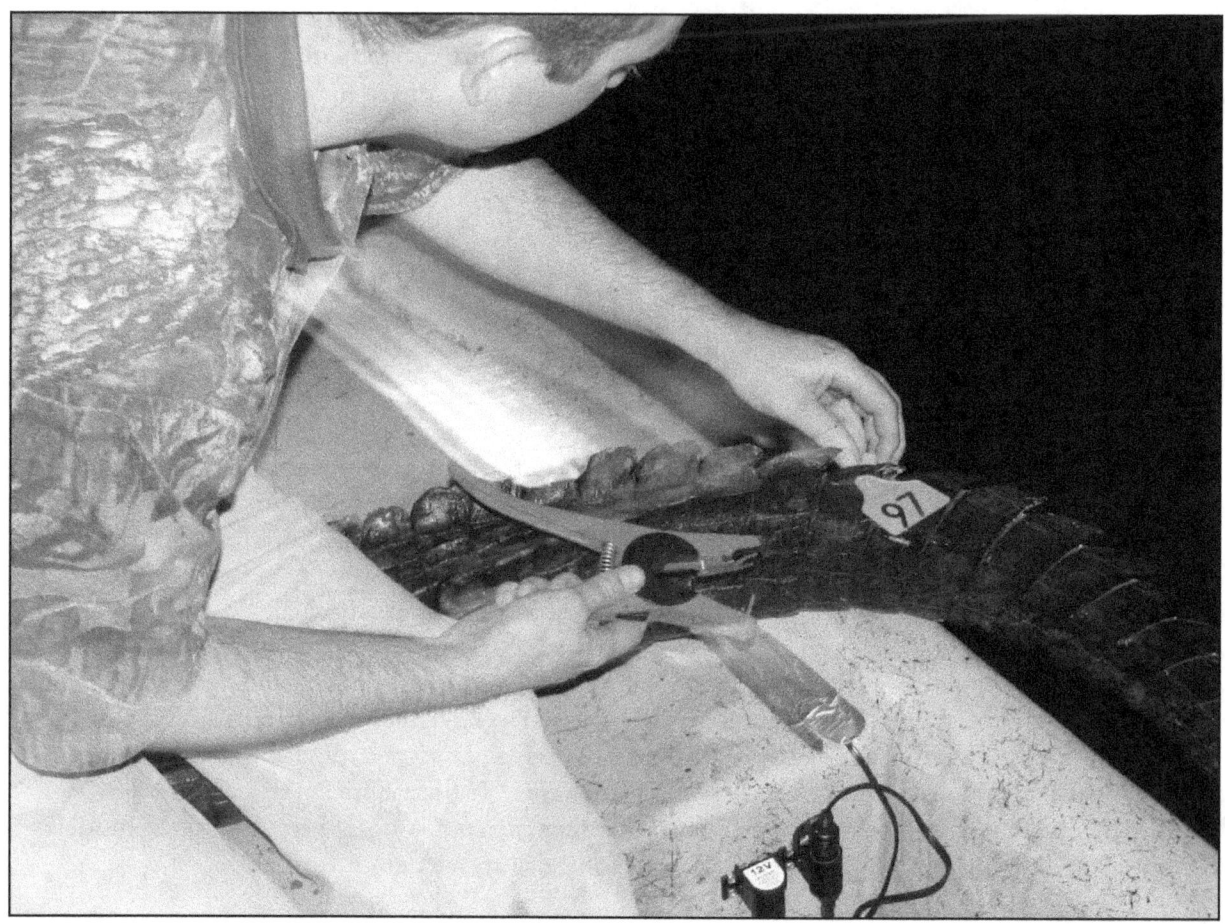

An American alligator (Alligator mississippiensis) is tagged with an individual ID tag on a raised tail scute by biologist Russ Walsh (Mississippi Wildlife, Fisheries and Parks) during a night capture exercise. Photo by Tim Hiller (used with permission).

Ch. 11

Recovery data[1]

"If it looks like a duck, and quacks like a duck, we have at least to consider the possibility that we have a small aquatic bird of the family Anatidae on our hands."
-- Douglas Adams

Questions to ponder:
- *How does recovery data differ from CJS-type data?*
- *Does sampling fish with anglers (fishermen) or sampling ducks with hunters bias the survival rates I am trying to estimate?*
- *How do I define the encounter rates that are critical to dead recovery data?*

Models for game species

Few biologists use models to estimate survival from marked animals that are subsequently recovered dead. But, those few biologists work on some important game species—usually waterfowl and fish. These species are typically harvested, at which time the mark (either a band/ring or a tag) is discovered and reported. The reason that few people use this type of data is that a biologist has to mark a large cohort of animals to ensure that even a few harvested animals have tags. Thus, "recovery" programs usually involve large efforts—such as the waterfowl banding effort by state wildlife agencies in the United States, the US Fish and Wildlife Service, and the Canadian Wildlife Service.

In Chapter 10, we discussed the "apparent survival problem"—created when animals emigrate beyond the scope of the capture efforts. In such situations, animals that emigrate are considered "dead" to the model, as the CJS-type analysis cannot distinguish between emigration and mortality.

Recovery models typically have built-in characteristics that solve the apparent survival problem. For most situations, hunters and anglers (the "recoverers") exist at a larger spatial scale than the study. So, it is almost impossible for an animal to move beyond the reach of people who would report the animal, if it were harvested.

---

[1] *With thanks for content to Michael Conroy, Gary White, William Clark, and Evan Cooch*

Brownie models

Cavell Brownie, with her important publication (Brownie et al. 1985), developed models that advanced the use of dead recoveries for estimation of survival from waterfowl. The context of these "Brownie" models should be remembered and explained: waterfowl are typically marked with bands, or rings, in late summer (August and early September) in the northern United States and central Canada. Waterfowl then migrate south and are harvested by hunters during migration—in September-January, typically. The next spring, the birds return north and are once again targeted for marking operations by biologists in August (Figure 11.1).

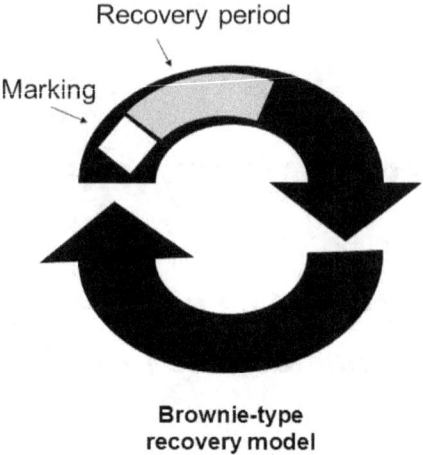

Figure 11.1: *Annual cycle showing timing of marking and recovery of animals under the Brownie-type recovery model.*

The expected band returns in year i from releases of a cohort, R_i, in the same year is

$$E(r_{ij}) = R_i f_j$$

Thus, **direct recoveries** (recoveries that occur immediately after banding/ringing, Figure 11.1) are a function of only the sample size released, R_i, and the probability of recovery, f_i. Note that survival, S_i, is not a part of the formula, because we assume that harvest happens immediately after marking, and no birds die between marking and harvest. However, survival is important for **indirect recoveries**, which must survive ≥1 year after marking before being recovered. Thus, to estimate the expected tag returns of indirect recoveries in <u>any year j</u> after releases in year i:

$$E(r_{ij}) = R_i \prod_{h=i}^{j-1} S_h f_j$$

When a bird is released with a band, or ring, we can imagine three possible fates (Figure 11.2): (1) the bird may survive the year, (2) the bird may be killed by a hunter, and (3) the bird may die

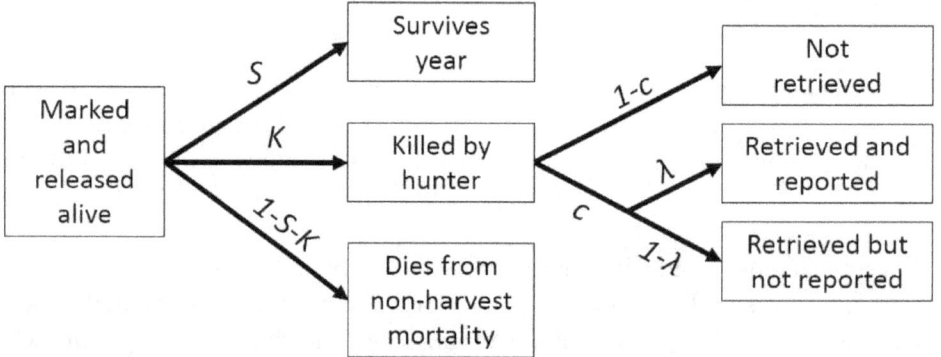

Figure 11.2: *Possible changes in status for animals after being marked and released under general recovery model theory. Animals killed by hunters have further possibilities for retrieval and reporting of data. Probabilities are defined in the text and are shown on the transitions in this figure.*

of non-hunting causes. We can identify, then, two parameters: S, the probability of surviving one year, and K, the probability of harvest by a hunter during one year. When K is applied to all birds in the population, waterfowl biologists also define K as the **harvest rate**: the proportion of the population to be affected by harvest in a given year (Figure 11.2).

However, band recovery is definitely not a known-fate process—every bird that is banded and harvested is not necessarily reported. In fact, some birds are shot and never found by the hunter, so the birds are not **retrieved** to the hunter's bag. And, hunters who retrieve harvested birds with marks may not **report** the data to a biologist. Thus, we need to identify some parameters to represent the probability of being retrieved (given harvest), c, and the probability of being reported ("reporting rate", given harvest and retrieval), λ (Figure 11.2).

Brownie model structure

Brownie et al. (1985) provided a structure to estimate two parameters: S, the probability of surviving one year, and f, the probability of recovery. **Recovery**, then, becomes the product of $K*c*\lambda$ (Figures 11.2 and 11.3). That is, to be "recovered", a bird must first by harvested, and then retrieved, and then reported. A failure at any point along this chain of actions results in a lack of recovery (and therefore data).

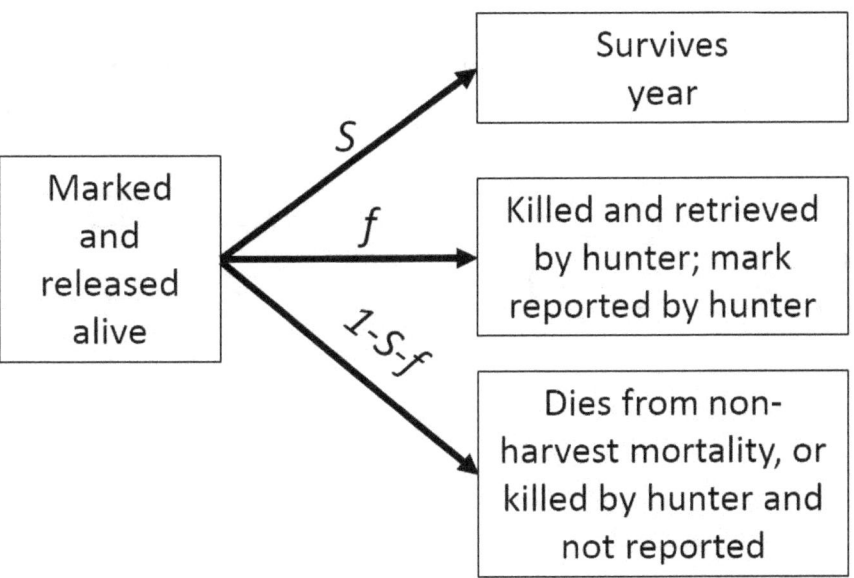

Figure 11.3: *Possible changes in status for animals after being marked and released using Brownie-type recovery models with binomial events of survival (S) and recovery (f). The reader should note that this figure is a simplification of Figure 11.2.*

We can define the probabilities of being released in a given year, i, as part of a cohort of banded animals, R_i, as:

Banded, released	Probability of recovery in Year		
	1	2	3
R_1	f_1	$S_1 f_2$	$S_1 S_2 f_3$
R_2		f_2	$S_2 f_3$
R_3			f_3

Again, we stress the unique feature of Brownie-type recovery models that were built for waterfowl—first-year recoveries, gathered immediately after marking, do not require survival before encounter.

Estimating survival from recoveries

The data available, then, for the estimation of survival from banded waterfowl can be summarized as follows, where m_{ij} represents the number of animals recovered in year j that were marked in year i.

Banded, released	Recovered in Year		
	1	2	3
R_1	m_{11}	m_{12}	m_{13}
R_2		m_{22}	m_{23}
R_3			m_{33}

The logic to the estimation of survival for recovered animals is that the recovery rate should be the same for all cohorts of released animals (R_1, R_2, R_3, etc.) in a given year. That is, ducks banded this year or last year are experiencing the same hunter pressure (the same harvest rate, Figure. 11.2) and retrieval and reporting levels.

As in Chapter 10 and the CJS survival estimator, our survival estimator for recovery-type data, where M_{ij} represents the number of marked animals still alive in the population, is

$$\hat{S}_1 = \frac{\hat{M}_{12}}{R_1}$$

And, as with CJS, the problem is that our open population makes it impossible to directly know M_{ij}. So, we must estimate M_{ij}.

As an example, we could postulate that if recovery rate of our marked animals was 100%, then M_{12} would be equal to m_{12}. But, we know recovery rate is not 100%. But, it will be possible for us to estimate the recovery rate, so we can use the recovery rate to estimate M_{12} (and by extension, any M_{ij}) as:

$$\hat{M}_{12} = \frac{m_{12}}{f_2}$$

How do we estimate the recovery rate? It is simply the proportion of the banded cohort that is recovered in a given time period:

$$\hat{f}_2 = \frac{m_{22}}{R_2}$$

Then, we can provide the following estimator for survival for year 1 by substituting and doing a bit of algebra from the above equations:

$$\hat{S}_1 = \frac{m_{12} R_2}{m_{22} R_1}$$

The structure of the dead recovery model has **assumptions**, which are:
- Survival estimates are not biased by the method of retrieval (hunting).
- Marked animals are representative of target population.
- Data is properly recorded, no mark loss.
- Survival is not affected by tagging (hunter/angler doesn't make decision to 'sample' because of tag).
- Fates of tagged animals are independent.

The first assumption may bear some explaining, as this concept can be confusing.

First, it is certainly true that animals die when hunters shoot them (Fig. 11.4), or when anglers catch them (assuming they are not returned to the lake). But, the act of shooting or angling did not take place with the purpose of obtaining the tags—the tags are found (in secondary fashion) by hunters who were hunting for food and recreation. Therefore, we use anglers and hunters to sample the population at time i, and we assume that the

Figure 11.4. Hunters are used to sample waterfowl in the North American Bird Banding Program, such as the northern pintail (Anas acuta) in hand here. IDs on bands are reported to the Bird Banding Laboratory in Laurel, Maryland, USA (photo provided by Stephen Siddons and used with permission).

retrieval of tags does not alter a harvest level that is in the process of being established by other forces.

Second, survival rates may indeed be a function of hunting/angling, as a source of mortality. But, again, the retrieval of tags does not bias that source of mortality, because we use hunters and anglers who are already in the field. If, for example, we sent 10,000 biologists out to harvest ducks for the sole purpose of obtaining more bands, we could be violating the first assumption, above.

And last, we must emphasize that survival and exploitation rate (harvest) are not necessarily linked—it is possible for the probability of harvest, K, to increase in a given year while survival also increases because of good weather or low populations of predators. The potential effects of density dependent over-winter survival (a mechanism for compensatory harvest mortality) also provide the opportunity for annual probabilities of S and K to be unrelated.

Seber's model structure

Earlier in this chapter, we stressed the unique feature of Brownie-type recovery models that were built for waterfowl—first-year recoveries do not require survival before encounter. This can be an important feature to remember if you are adapting Brownie-type models for use in other contexts—perhaps for capture of fish by anglers, for example. If the harvest takes place immediately following marking, the Brownie-type models should be suitable for you. But, if recovery takes place for example, 5-6 months after the initial marking, or right before the next marking event, the Brownie-type models will need adjustment to fit your situation—because your animals will have had to survive long periods before harvest.

Seber (1970) provided a more generalized recovery model that may be useful when recovery does not occur immediately after harvest. In this model, recovery is assumed to happen any time before the next time period (the next mark/release; Figure 11.5).

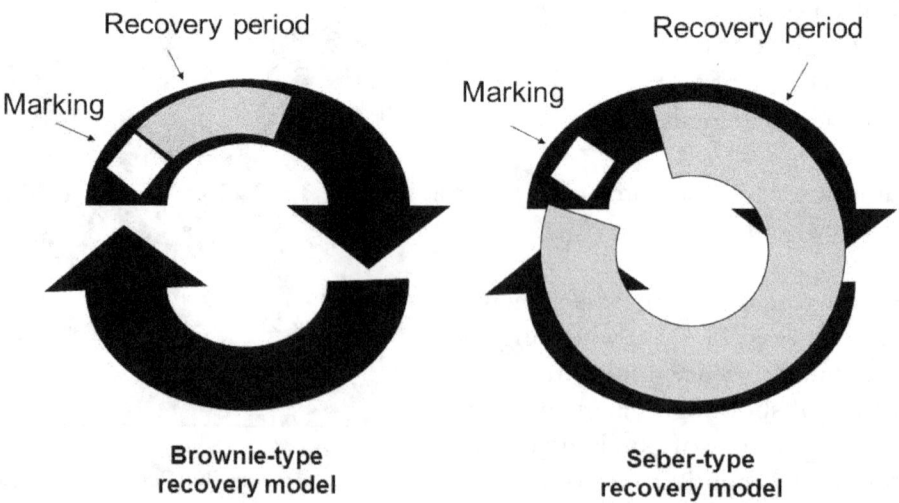

Figure 11.5: *Comparison of annual cycle of marking and recovery events in Brownie-type and Seber-type recovery models.*

Seber model structure

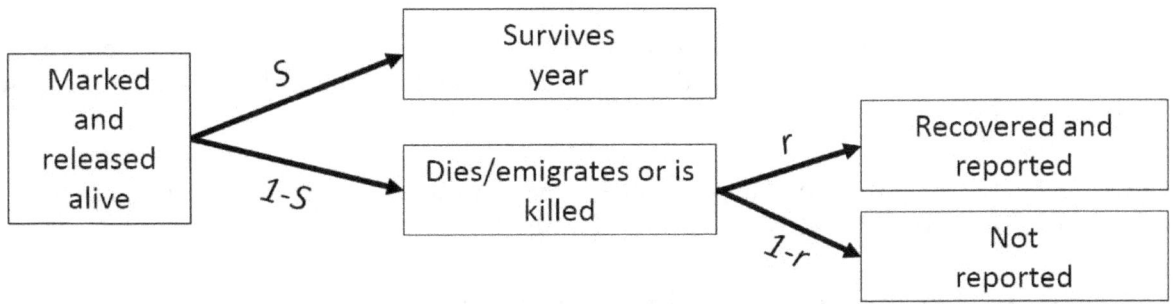

Figure 11.6: *Possible changes in status for animals after being marked and released using Seber-type recovery models. Animals that do not survive (S) have a further alternative for recovery/reporting (r).*

The difference between the structures of Brownie- and Seber-type recovery models can be seen in Figures 11.3 and 11.6. The contrast is the probability of being released as a member of cohort R_1 and then subsequently being recovered in either year 1, 2, or 3. Seber-type model structure does not allow recovery, r, in the same year as release, while it is a key feature of Brownie-type models. Both model structures may be useful for different situations if you have recovery data to use for survival estimation. Because Seber's structure is more generalizable, it may also be extended to non-harvest situations where animals are found dead or recovered in other ways.

Banded, released	Probability of recovery in Year		
	1	2	3
Brownie structure:			
R_1	f_1	$S_1 f_2$	$S_1 S_2 f_3$
Seber structure:			
R_1		$S_1 r_2$	$S_1 S_2 r_3$

Conclusion

Recovery data is limited to unique circumstances where large numbers of animals are released with tags and subsequently recovered (dead) by hunters or anglers. This type of data provides important information to manage game species, such as waterfowl, gamebirds, and harvested fish. Two model structures are available—Brownie or Seber—depending on the timing of the recovering following marking.

References

Brownie, C., D. R. Anderson, K. P. Burnham, and D. S. Robson. 1985. Statistical inference from band recovery data -- a handbook, 2nd ed. U. S. Fish and Wildlife Service Research. Publication 156, Washington, D. C. 305pp.

Seber, G. A. F. 1970. Estimating time-specific survival and reporting rates for adult birds from band returns. Biometrika 57: 313-318.

For more information on topics in this chapter

Conroy, M. J., and J. P. Carroll. 2009. Quantitative Conservation of Vertebrates. Wiley-Blackwell: Sussex, UK.

Cooch, E., and G. White. 2014. Chapter 8: 'Dead' recovery models. *In* Program MARK: a gentle introduction, 12th edition, Cooch, E. and G. White, eds. Online: http://www.phidot.org/software/mark/docs/book/pdf/chap8.pdf

Williams, B. K., J. D. Nichols, and M. J. Conroy. 2002. Analysis and management of animal populations. Academic Press, San Diego.

Citing this primer

Powell, L. A., and G. A. Gale. 2015. Estimation of Parameters for Animal Populations: a primer for the rest of us. Caught Napping Publications: Lincoln, NE.

Ch. 12

Multi-state models[1]

"Life is pleasant. Death is peaceful. It's the transition that's troublesome."
-- Isaac Asimov

Questions to ponder:
- *How do multi-state models build on CJS-type models?*
- *How much data do I need for a multi-state model?*
- *Can I have too many "states" in a multi-state model?*
- *Do multi-state models remove the problem of apparent survival found in CJS models?*

An adjustment to CJS-type models

We discussed CJS-type models to estimate survival in Chapter 10. The standard CJS-type study is completed at one study site, where animals are captured, marked, and recaptured or resighted over time. In Chapter 10, we discussed the problem of **apparent survival**—because CJS-type models cannot distinguish between animals that leave the study site and those that die. As you read Chapter 10, you might have thought to yourself, "It would be nice if we could estimate the probability of emigration." If you had those thoughts, you are ready to think about multi-state models.

Multi-state models are often used to study populations that exist at more than one geographic "state", or area. An individual can only exist in one state at one time. And, it must move between the states to **transition,** or change states. We will also see that multi-state models can be used in innovative ways to study transitions between other "states" for animals—for example, multi-state models could be used to model the probability of nesting and not-nesting during the breeding season. But, for now, we will use geographic states to introduce the model structure.

Multi-state models are a modification to the CJS model structure. We will retain our two initial CJS-type parameters: S, the probability of survival over a time period, and p, the probability of

---

[1] *With thanks for content to David Smith, Max Post van der Burg, John Carroll, Michael Conroy, and Gary White*

encounter (capture or resighting). And, we will add a parameter, ψ, the probability of movement.

The focus on states will also necessitate that our notation will become more complex—because we need to indicate time periods as well as states for our parameters. So, to fully notate our parameters, we will use:

S_i^r : probability of survival in state r during time i

p_i^r : probability of encounter in state r during time i

ψ_i^{rs} : probability of moving from state r to state s during time i

Our capture histories will become more interesting as well—can you guess what is happening to this animal moving in a 3-state system (Figure 12.1) based on a modification to the typical CJS-type capture history?

0A00B0BB0AC

The capture history indicates the animal was first tagged in area A during time 2. It was next seen during time 5 in area B, where it was seen again at time 7 and 8. The animal is known to have moved back to area A by time 10, and then immediately seen in area C in time 11. We note the importance of the fact that *we have no idea where the animal was* during periods 1, 3, 4, 6, and 9.

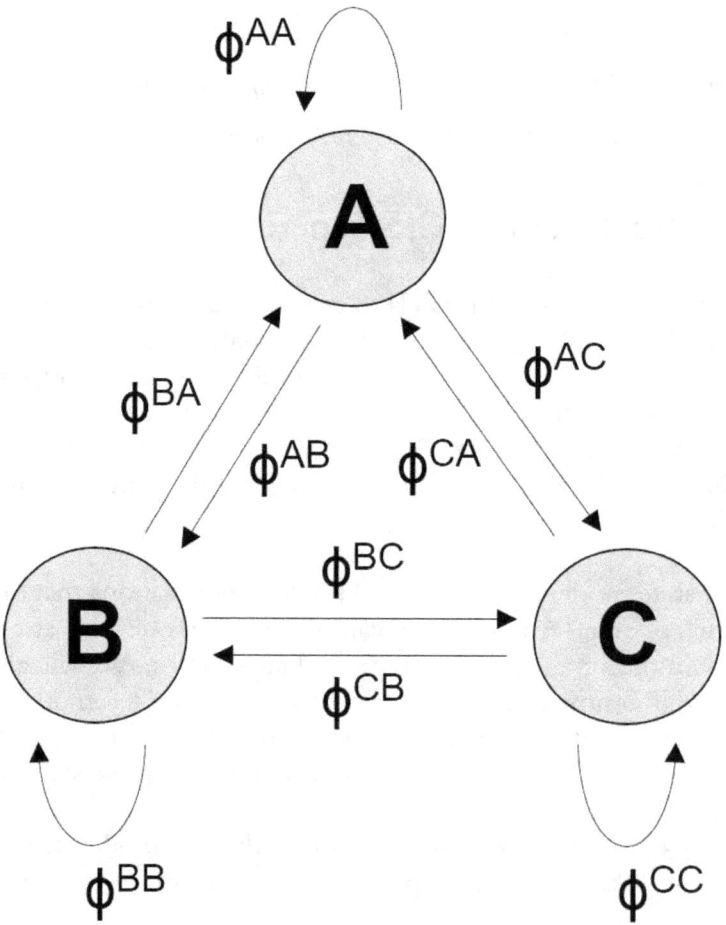

Figure 12.1: *Potential transitions in a simple system with three states, A, B, and C. For simplicity, we show the transitions, ϕ^{rs}, as the product of the probabilities of survival, S^r, and movement, ψ^s. The multi-state model structure also includes the probability of capture, p^r, in each state.*

We can examine the structure of multi-state models by writing out the probability statements associated with some example capture histories for a two-state system (with only states A and B).

The probability statements are straightforward when an animal is seen in all time periods, such as Animal #1 in the table at right. Given capture in time period 1 in state A, the probability of being captured again in state A

Animal	Capture history	Probability statement (*given initial capture*)
#1	AAB	$\phi^{AA} p^A \phi^{AB} p^B$
#2	A0B	$\phi^{AA}(1-p^A)\phi^{AB} p^B +$ $\phi^{AB}(1-p^B)\phi^{BB} p^B$

during time period 2 is: $\phi^{AA} p^A$, where ϕ^{AA} is the product of the probability of surviving from time period 1 to time period 2 in state A, S^A, and the probability of moving from state A to state A, ψ^{AA}. And, of course, the animal has to be captured in state A, as well (p^A). Now, the animal is in state A in time 2, and the probability of being captured in state B in time 3 is $\phi^{AB} p^B$. Then, ϕ^{AB} is the product of probability of surviving from time period 2 to time period 3 in state A, S^A, and the probability of moving from state A to state B, ψ^{AB}. And, the animal has to be captured in state B (p^B). If all of those binomial events occur (survival, movement, and capture), the animal will have a capture history of AAB and the probability of such a capture history can be stated as shown in the table above: $\phi^{AA} p^A \phi^{AB} p^B$.

But, when an animal is not captured in a time period, such as Animal #2 during time period 2, we must account for <u>all</u> potential paths in our probability statement. Therefore, we have to add the probability for all potential paths together—here, we know that Animal #2 was alive in time period 2 (because it was seen alive in time 3), but we do not know if it was in area A or area B during time 2. So, we include both possibilities. Using the details for Animal #1, we can see that we can represent the possibility of moving from A to A to B as: $\phi^{AA}(1-p^A)\phi^{AB} p^B$. And, we can represent the possibility of moving from A to B to B as: $\phi^{AB}(1-p^B)\phi^{BB} p^B$. In both cases, the animal is not captured during time period 2, so we use "1-p" to describe the probability of not being captured. And, because the animal does not do both possibilities (it is "either/or"), we get a complete statement of: $\phi^{AA}(1-p^A)\phi^{AB} p^B + \phi^{AB}(1-p^B)\phi^{BB} p^B$.

You try it!

First, to make this example a little more realistic, we can assign some values to the probabilities. Let's assume that capture probability (p) is 20% in state A and 15% in state B. Let's assume survival probability (S) is 60% in state A, and the probability of movement for animals in state A is as follows: A to A: 80%, A to B: 20% (*notice that an animal in state A has to go somewhere if it stays alive, so these values have to add up to 100%!*). Given those estimates for these parameters (and given the initial capture in time 1 in A), what is the probability of having a capture history AAB?

Second, can you write out the possibilities, for the same two-state system as above (A and B), for an animal that was only seen once: A00? To make it simpler, assume the animal did not die. Where could it have been in time 2? Time 3? *See end of chapter for answer.*

You must survive before you move

Above, we used ϕ^{rs}, as the product of S^r and ψ^{rs}. Here, we emphasize the logic used in multi-state models, with regard to survival and movement.

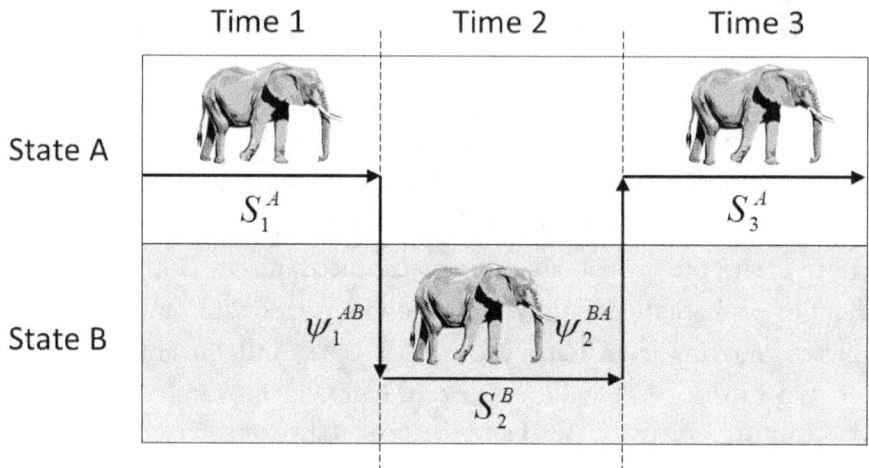

Figure 12.2: *The logical order of survival and movements used in multi-state model structure. Animals must first survive (S) through a time period (1, 2, or 3, here) in a state (A or B, here) before moving (ψ) at the last moment in the time period to another state. Here, an elephant transitions from state A to state B and back to state A. Capture probabilities (p) are not shown.*

To make multi-state models work properly, a basic assumption is needed—animals tagged or recaptured/resighted in a given state must remain in that state until the very end of the current time period. **That is, animals must survive before they move.** This is a key concept to multi-state models. And, it is why we use the strata identifiers that we do (Figure 12.2). The transition from strata *r* to strata *s* first involves surviving in strata *r* and then moving to strata *s* at the last possible moment during time *i*, before *i+1*.

$$\phi_i^{rs} = S_i^r \psi_i^{rs}$$

The main reason for the assumption is that we must give the 'critters' (animal of interest) a 'home base' for a time period so that we can estimate state-specific (or strata-specific) survival. The structure also provides the key to eliminating the problem of 'apparent survival' (at least in concept, *stay tuned!* [see our word of caution below])—because we have information to estimate the movement in and out of a strata, which effectively strips "emigration" away from "mortality" and our survival becomes a 'more pure' "survival".

An embarrassment of riches: too many parameters?

Earlier in this chapter, you practiced writing probability statements for capture histories from a study in a simple two-state system (states A and B) over three time periods. *Now, imagine the statement needed in a system with many states for an animal with the capture history:* A00DEF00BQ Don't worry, we won't write them all out! But, you may start to see a challenge for multi-state models—there are many, many parameters needed. As we have mentioned before, as you add parameters, you need to add data to support those parameters. If you don't find tagged animals moving between sites (e.g., site M to Q), you don't have any data to inform the value for the probability of that transition. And, your sample size is split among all of the areas—necessitating the tagging of many more animals to support the same level of inference compared to a typical CJS-type study in one area. For this reason, we encourage you to consider minimizing the number of strata, if you can, if you are interested in a multi-state study. Models with fewer strata can lead to more precise estimates from the same set of data. If you cannot reduce the number of strata needed, you should start writing grant proposals to provide funds for more workers and equipment at all of the sites!

Further, as we saw with the CJS model structure, not all parameters may be identifiable (see Chapter 10). In time-dependent, multi-state models, the probabilities of survival, movement, and recapture are confounded in the final time period and are not estimable. In addition, as suggested above, sparse data at any point may render some parameters inestimable.

To put a final point on our emphasis on the "problem of parameters" of multi-state models, here is a table with the number of parameters to be estimated in a generalized model, for each time period:

Number of states, r, in model structure	Number of state-specific parameters to be estimated for each time period, i			
	p_i^r	S_i^r	ψ_i^{rs}	Total
2	2	2	2	6
3	3	3	6	12
4	4	4	12	20
5	5	5	20	30
8	8	8	56	72
10	10	10	90	110

Multi-state model assumptions

In addition to the assumptions of CJS model structure, multi-state models have the following assumptions:

- Data are recorded without error with respect to the stratum of recovery, resighting, or recapture.
- All animals in stratum r at time i have equal movement probabilities.
- Movement probability and recapture probability do not depend on the past history of the animal.
- All animals make transitions at the same time during the interval.

The third assumption, above, is important to remember—multi-state models, in their basic form, have no "memory". So, an animals is no more likely to move than another, even if the animal has a documented history of changing back and forth between strata during previous time intervals. Movement transitions are estimated similar to survival estimates—by assessing a group of animals in a specific strata and determining their fate. So, the "busy" individual (one that moves around a lot) simply joins its strata-mates in providing a sample of animals from which we estimate a transition rate.

Think outside the strata

Multi-state models have been used primarily for the estimation of movement rates between geographic strata, but if you think of an animals' annual cycles as the transition between many different types of "states", you may see a new way to use multi-state models. For example, you might use the states to indicate breeding status, and the transitions indicate the probability of switching between breeding and non-breeding. Perhaps you are interested in the probability of being in breeding status during a given season. Or, you could view the states as reproductive 'age'. Some animals have less deterministic transitions from juvenile to adult.

So, you might use multi-state models to estimate the probability of the indeterminate transition from juvenile to breeding status in black rhinos (Law and Linklater 2013, Figure 12.3). A paper by Nichols and Kendall (1995) provides encouragement for this interesting application for multi-state models.

Figure 12.3: *A black rhino (Diceros bicornis), marked with ear notches, in a national park in Namibia. Photo by L. Powell.*

Final thoughts about apparent survival: caution

Although multi-state models may solve much of the problem of apparent survival encountered with the use of only one strata and CJS-type models, we must emphasize that the problem is not completely solved. If, at any time, one of your study animals emigrates outside of your multi-state system, the strata-specific survival estimates may be negatively biased. Certainly, it is rarely possible to completely surround your study animal by including all possible geographic strata (Figure 12.4). So, the problem of apparent survival remains, although the negative bias should be greatly reduced through the estimation of movement probabilities in a multi-state model structure.

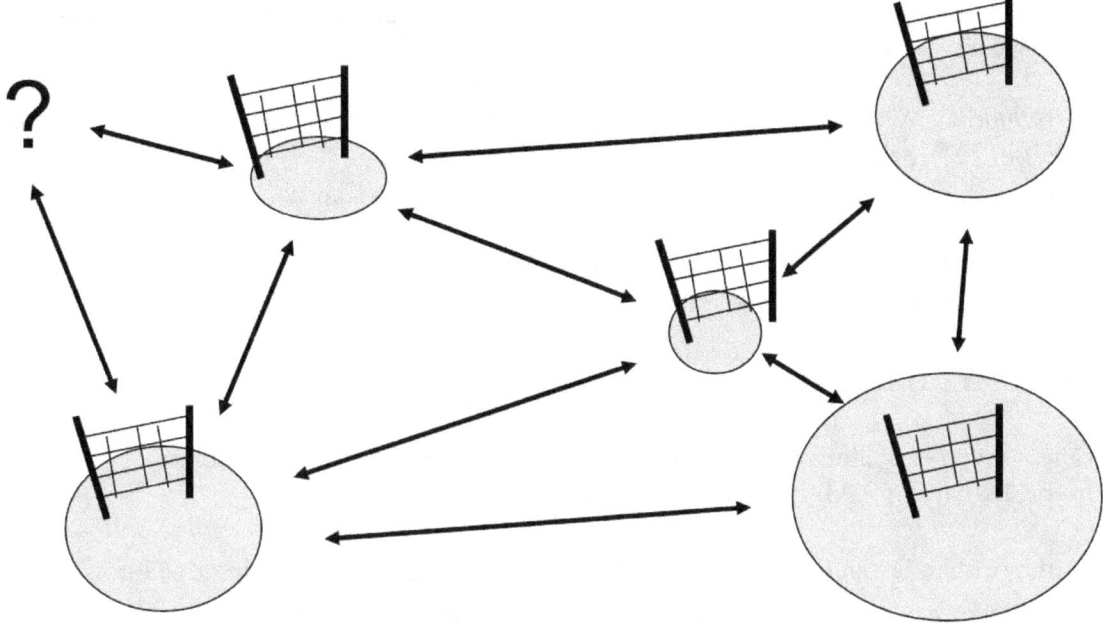

Figure 12.4: *A hypothetical multi-state study system with 5 locations where animals can be located and potentially captured (with mist nets, as shown, or by other methods such as traps or resighting) as they move through the system. However, if animals move outside of the system (noted by "?"), there is no potential to recapture them. Some potential transitions are not shown for clarity.*

And, if you are using multi-state models to estimate probability of breeding transitions (e.g., non-breeding to breeding), rather than geographic transitions, for a study animal in a single location, the survival rates estimated will still be apparent survival rates.

Conclusion

Multi-state models are a modification to the CJS-type model structure, and we use multi-state models to estimate movement rates, survival rates, and encounter rates. Multi-state models are parameter-rich and their use requires forethought and more support in the field to provide data that can be used to estimate the large number of parameters. These models can be used to estimate transitions between geographic strata, or other types of states that can be identified for cohorts of animals.

References

Nichols, J. D. and W. L. Kendall, W. L. 1995. The use of multi-state capture-recapture models to address questions in evolutionary ecology. Journal of Applied Statistics 22: 835-846.

Law, P. R. and W. L. Linklater, W. L. 2013. Black rhinoceros demography should be stage, not age, based. African Journal of Ecology 52: 571-573.

For more information on topics in this chapter

Conroy, M. J., and J. P. Carroll. 2009. Quantitative Conservation of Vertebrates. Wiley-Blackwell: Sussex, UK.

Cooch, E., and G. White. 2014. Chapter 10: Multi-state models... *In* Program MARK: a gentle introduction, 12th edition, Cooch, E. and G. White, eds. Online: http://www.phidot.org/software/mark/docs/book/pdf/chap10.pdf

Williams, B. K., J. D. Nichols, and M. J. Conroy. 2002. Analysis and management of animal populations. Academic Press, San Diego.

Answers: You try it!

Part One: Given the values for the parameters provided, what is the probability of having a capture probability of AAB in a two-state system (states A and B, only)?

We use the probability statement given in the table: $\phi^{AA}p^A\phi^{AB}p^B$. Breaking this into steps by time period, we first calculate the probability of $\phi^{AA}p^A$ as: 0.6*0.8*0.2 = 0.096. That is, an animal in A has a 9.6% chance of surviving in A, staying in A, and being recaptured in A in the next time period. Then, we calculated $\phi^{AB}p^B$ = 0.6*0.2*0.15 = 0.018. Thus, an animal in A has a 1.8% chance of surviving in A, moving to B, and being recaptured in B. Multiplying these probabilities together, we arrive at our complete probability for having a capture history of AAB: 0.096*0.018 = 0.002, or 0.2%. If you released 1000 marked animals in state A during time period 1, you would predict that only 2 of them would have a capture history AAB at the end of the study. *Does that sound low to you?* Remember that your survival rate was 60%--you lose 40% of your animals to mortality every time period. Second, the capture history includes a "B", and the probability of leaving A for B is very low (0.20), so most animals that start in A will stay in A. Last, your capture probability was very low as well (0.20 or 015, depending on the state). So, 80% of the surviving animals will not be captured in a given time period. *This scenario emphasizes why it is important to mark as many animals as possible!*

Part Two: In a two-state system (A and B), what are the possible movements for an animal with a multi-state capture history of A00?

Answer: This animal might have stayed in A during the next two time periods, it might have stayed in A during the next time period and then transitioned to B in time 3, or it might have transitioned to B in time 2 and then come back to A in time 3, or it might have transitioned to B in time 2 and stayed there.

Thus, the possibilities (if it survived and if we had perfect knowledge of its locations) are:

AAA, or AAB or ABA or ABB

We have to account for all four of the possibilities. Therefore, the probability that an animal will have a capture history of A00 is:

$$\phi^{AA}(1-p^A)\phi^{AA}(1-p^A) + \phi^{AA}(1-p^A)\phi^{AB}(1-p^B) + \phi^{AB}(1-p^B)\phi^{BA}(1-p^A) + \phi^{AB}(1-p^B)\phi^{BB}(1-p^B)$$

Citing this primer

Powell, L. A., and G. A. Gale. 2015. Estimation of Parameters for Animal Populations: a primer for the rest of us. Caught Napping Publications: Lincoln, NE.

A field isoflurance unit is used to anesthetize a Virginia opossum (Didelphis virginiana) before a radio-telemetry collar is attached in a wetland system in Nebraska, USA. Mesopredators, such as the opossum, move between wetland patches, and transition probabilities may vary by species.

Ch. 13

Robust design[1]

"The investigator should have a robust faith—and yet not believe."
-- Claude Bernard

Questions to ponder:
- *Why is this model structure called 'the robust design'?*
- *How are closed and open models combined in this model structure?*
- *How can the model estimate emigration?*
- *How can I estimate the size of a super-population with the parameter estimates from my analyses?*

The ways of innovation

Advances in methods for parameter estimation allow new perspectives on the dynamics of populations. The 'spark' required for these advances requires basic understanding of the probability-based models that are present and some innovation on the part of the scientist. The robust design approach to mark-recapture analyses is an example of innovation and, simply, a good idea. Here is a quote from Kenneth Pollock, the 'father' of robust design as he explains the inspiration for his initial work (Pollock 1982):

> *"During the preparation of a review of capture-recapture methods….I realized that statisticians have drawn a sharp distinction between closed and open population models that is perhaps rather artificial. Here I describe a design for long-term studies that is robust to het-erogeneity and/or trap response. It allows an analysis that uses methodology from closed and open population models."*

The model structure that Pollock provided was innovative, and it was also prone to quick use by ecologists because it did not require (in most cases) a new approach to sampling. In fact, Pollock's approach uses the annual sampling approaches that are common to many fisheries or wildlife monitoring schemes—an annual repetition of several capture events that occur in a relatively short period of time. The robust design is so-named, because Pollock suggested that it was "robust" to the problem of unequal probability of capture—something that was assumed in contemporary models of his day for parameter estimation from either open or closed populations. Thus, the problem of trap-shyness or sampling efforts that varied through time could be alleviated with the new model structure.

---

[1] *With thanks for content to William Kendall, James Nichols, and Michael Conroy*

Pollock initially conceived his model as being useful to estimate a new parameter—birth rate. In subsequent work, Nichols and Pollock (1990) provided a mechanism to modify the robust design to estimate immigration. Eventually, Kendall et al. (1997) modified the method to allow estimation of emigration as well as immigration. The Robust design has undergone multiple recent innovations, including combining it with multi-state model structures, but we will focus on the basic robust design to provide a foundation to understand more complicated model structures covered in more advanced texts.

Robust design basics

In a typical CJS-type model (see Chapter 10), we assume that an animal with the capture history of 101 was at the study site during time 2, but was not captured. That is, the structure of CJS-type models does not allow for temporary emigration. We can imagine, of course, many situations where that assumption does not hold, and our estimate for the encounter probability is then biased low. Animals have home ranges or movements that take them, temporarily away from the area that we have selected as our study site. It is very conceivable that animals spend some time, during our study, away from the study site.

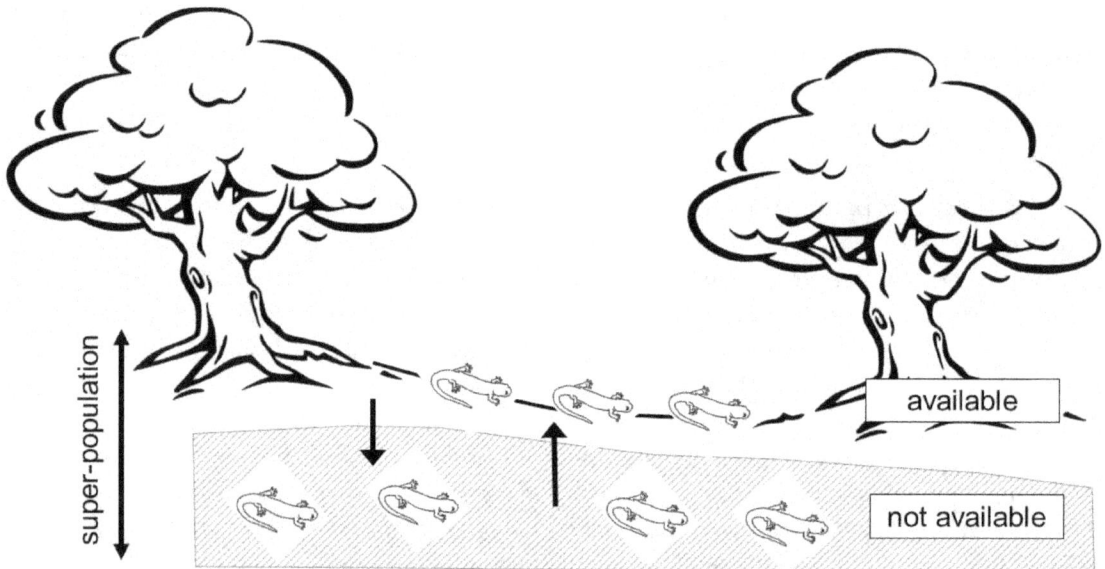

Figure 13.1: *Depiction of the super-population concept after Bailey et al. (2004). Terrestrial salamanders are available for counting by surveyors only when they are on the surface of the forest floor. The super-population consists of both available and unavailable animals.*

An example provided by Bailey et al. (2004) involves terrestrial salamanders (*Plethodontidae*) that temporarily emigrate away from the surface of the forest where biologists monitor them. The problem was that population size estimates were thought to be constantly lower than the actual population size, because only the salamanders on the surface were captured. And, in fact, Bailey et al.'s (2004) analysis confirmed that "significant proportions of terrestrial salamander populations are subterranean." In fact, the biologists estimated that 87% of the population was unavailable, and thus the animals that could be sampled on the surface was a small fraction of the entire population. The robust design is a good approach when a large portion of the population is not available for encounter during a given sample period.

In robust design terms, we refer to the set of animals that may be sampled at some point during our study as the **super-population**. The super-population, by definition, is larger than the subset of the population that is "available" for encounter during a short period of study because other individuals are not available. But, those not-available individuals in the super-population may become available at some time in the future, or they were available at some time in the past.

When a biologist attempts to estimate "population size" for a population, they often are interested in the size of the super-population—not just the size of the small group of animals currently inhabiting their study site. And, this is a good reason to use the methods of robust design.

Combining open and closed structures

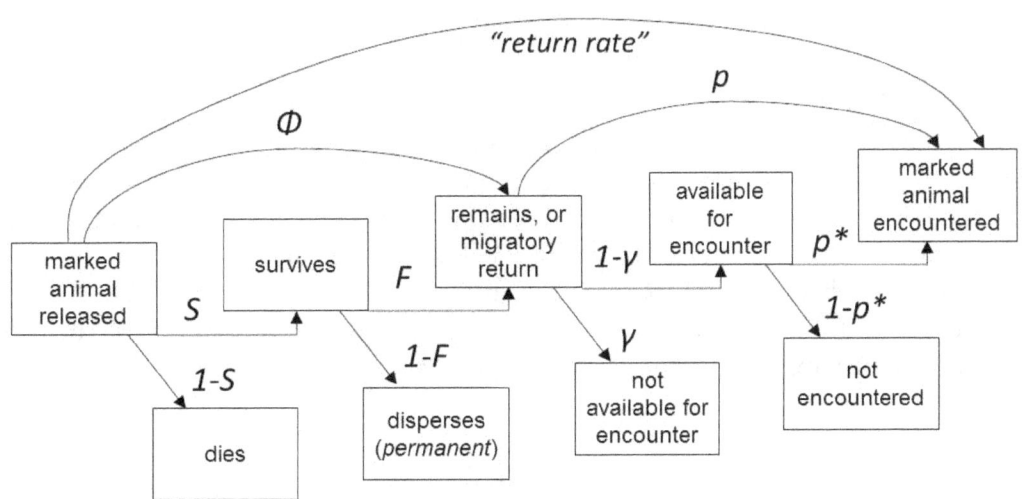

Figure 13.2: *Description of potential fates of a marked animal released in time i and potentially encountered again in time i+1 (after Kendall 2014).*

The robust design allows the estimation of many key parameters that describe dynamics of populations. It is nice to know population size and probability of survival, but in addition, knowledge of movement patterns is critical if we want to understand the complicated dynamics of a population.

Figure 13.2 shows all possible fates of an individual after it is marked. The animal is either going to survive or die during the next time period. If the animal survives, it may disperse from the super-population and never return—we label this **permanent emigration** and will distinguish it from **temporary emigration** later.

We previously discussed the permanent emigration problem when we described "apparent survival" in the Cormack-Jolly-Seber model structure. As we will soon see, the robust design is also prone to the apparent survival problem. That is, the basic model structure does not allow us to estimate **fidelity** (*F*), or permanent emigration (1-*F*). We will continue to distinguish "true" survival, *S*, and apparent survival, ϕ. Thus, the robust design model structure (e.g., Figure 13.2) will use ϕ as a reference to apparent survival.

If an animal remains on the study area (or returns to the study area after seasonal migration), it is possible that it may be **available** for encounter or it may be unavailable. Animals that are unavailable might be below ground (as in our salamander example) or outside of the boundaries of the area subject to netting or trapping. Terminology of the robust design defines the probability of being "not available" as **temporary emigration**, γ.

Last, if an animal is available, it may either be encountered (p^*, trapped, sighted, etc.) or not. And, we note that the CJS-type encounter probability, p, is a function of both availability (1-γ) and true encounter probability. In parallel fashion to "apparent survival", we might refer to CJS-type encounter probabilities, *p*, as "apparent encounter probability" because they do not account for the lack of availability. And, we can re-define *p* for robust design as the probability of encounter for a member of the super-population during a given time period. Thus, we see the relationship:

$$p = (1-\gamma)p^*$$

Before we leave the discussion of the structure behind the robust design model, we should also point out the relationship to "return rate" (discussed in Chapter 10). The crude "return rate" is usually defined as the proportion of animals released that return to a study site—and this has a deep history in waterfowl management in particular. Here, we see that **return rate** is actually the product of apparent survival, ϕ, and apparent encounter probability, *p* (Figure 13.2). And, each of these parameters is a function of other parameters: ϕ is a function of survival, *S*, and fidelity, *F*, and *p* is a function of temporary immigration, γ, and encounter probability, p^*. So, return rate, R, can be defined as

$$R = SF(1-\gamma)p^*$$

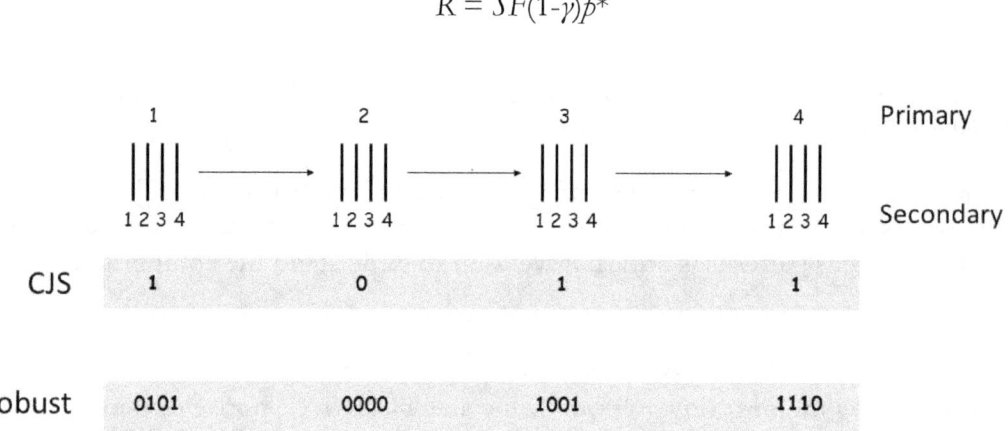

Figure 13.3: *The 'robust design' for mark-recapture including primary and secondary observation periods. Hypothetical capture histories are shown for all secondary periods under the robust design; capture success can be summarized into a simpler capture history for the primary periods used in Cormack-Jolly-Seber model structures.*

Model structure and capture histories

The basic robust design capture history, as described by Pollock (1982) is divided into "primary" and "secondary" sampling periods (Figure 13.3). The secondary sampling periods are considered closed—the time-frame, of course (see Chapter 8), depends on your study animal. But, it is within these secondary periods that we will be estimating closed-form population size, and we can obtain an estimate of N for each secondary period.

The primary periods are created by grouping the secondary periods (Figure 13.3). Modern software typically allow the length of time of the primary periods (the gaps between the secondary periods) to vary, although you must account for it in the model structure. The population is considered "open" between the primary periods.

You should be able to see that you could revert to a simple CJS-type analysis if you were to summarize the data from secondary periods into either encountered or not-encountered for each primary period. In this example, we would have a 4-occasion CJS-type capture history.

Similarly, we could conduct 4 separate closed-analyses of population size from our data. This is often done by biologists. However, Pollock's (1982) inspiration was to realize that more information could be obtained from a combined model structure—information about change in population size due to movements.

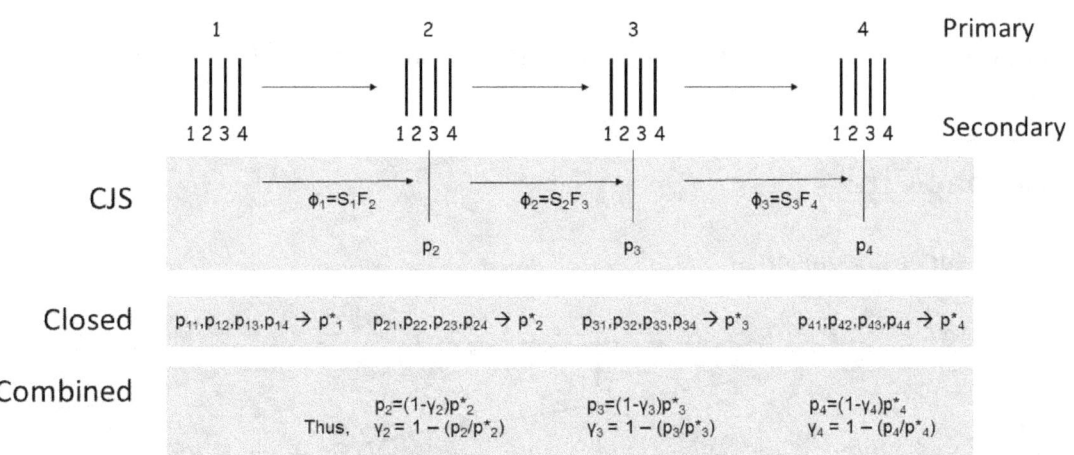

Figure 13.4: *Relationship of key parameters in the robust design (after Kendall 2014).*

What would this look like in the real world?

Imagine you had a 4-year study in which you conducted trapping of a lizard during only 5 days (of the same week in early July) each year. Certainly, we'd consider the lizard population to be closed during the 5-day period. Your primary and secondary periods might look like this:

Primary:	Year 1	Year 2	Year 3	Year 4	Year 5
Secondary:	1 2 3 4 5	1 2 3 4 5	1 2 3 4 5	1 2 3 4 5	1 2 3 4 5

How does the robust design estimate temporary emigration? It may help to clearly lay out the parameters (Figure 13.4) that are estimated by the open (CJS-type) and closed estimators, so that we can see the manner in which the combined information is used to estimate availability ($1-\gamma$, the complement of temporary emigration).

First, we can see that in the closed population model structure, we can obtain a time-specific estimate for p. Next, we can calculate the overall probability for capture, p^*, during the first primary period (Figure 13.4) as:

$$p_1^* = 1 - ((1-p_{11})(1-p_{12})(1-p_{13})(1-p_{14}))$$

> NOTE: We use the subscripts, p_{ij}, to indicate the primary period, i, and the occasion of the secondary sampling period, j.

Our model structure also allows the estimation of the "apparent encounter probability", p_i, for the primary period from the open-portion of the model structure (CJS; Figure 13.4). And, we can see that if all animals were available ($\gamma = 0$, or no temporary emigration), p_i should equal p^*. Because of this, we can estimate γ as a function of the relative difference of p and p^*:

$$\gamma_i = 1 - \left(\frac{p_i}{p_i^*}\right)$$

This is a fairly intuitive, and very important concept that is at the heart of the robust design. It's that simple!

Out, and back in?

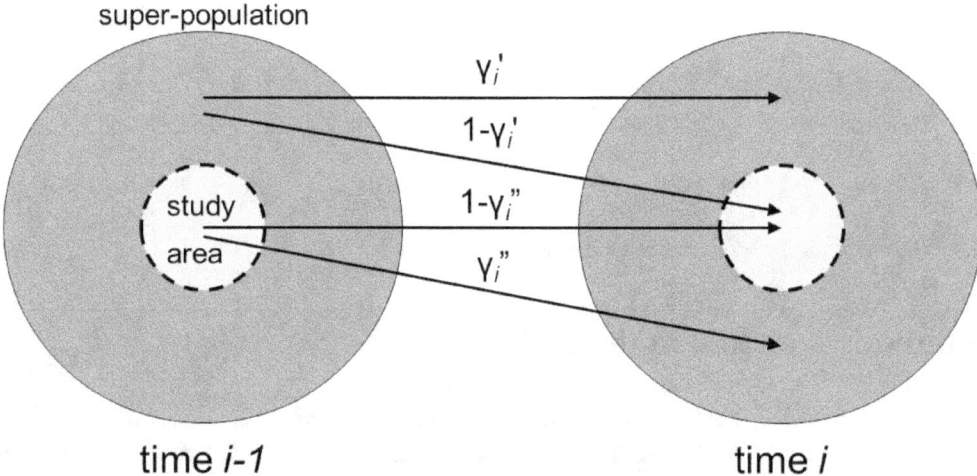

Figure 13.5: *Relationships between γ' and γ''. The super-population is represented by the larger circle, and the portion of the super-population available for encounter in the study area is represented by the inner circle. Animals temporarily immigrated with a probability of γ'' and are then not available for encounter. Animals return to the study area with probability of $1-\gamma'$.*

Kendall et al. (1997) provided one additional modification to the basic robust design, which forms the structure of the robust design model that is now typically available in most software packages. Kendall et al. (1997) realized that we could also be interested, over time, in what happens to animals once they leave—that is, we're interested in temporary emigration, but we are also interested in the rate of immigration back into our study site (Figure 13.5).

So, we now have a modification to the simple γ in our previous descriptions:

> γ': for animals that are away from the study site (but in the super-population) the probability of remaining away from the study site.
>
> Defined by Kendall et al. (1997) as the probability of being off the study area, unavailable for capture during primary trapping session i, given that the animal was not present on the study area during primary trapping session $i\text{-}1$, and survives to trapping session i.
>
> γ'': for animals on the study site, the probability of moving away from the study site.
>
> Defined by Kendall et al. (1997) as the probability of being off the study area, unavailable for capture during primary trapping session i, given that the animal was present on the study area during primary trapping session $i\text{-}1$, and survives to trapping session i.

Movement models

Current analyses with the robust design model, subsequent to Kendall et al. (1997), typically consider three movement models as competing models of reality:

> **No movement (null):** $\gamma' = \gamma'' = 0$; all animals <u>available</u> for capture.
>
> **Random:** $\gamma' = \gamma''$; chance of being 'out' is the same for 'in' and 'out' animals (that is, every animal in the super-population has the same probability of being outside the study area during the next time period)
>
> **Markovian:** the animal's location in a previous time period affects γ' and γ''; chance of being 'out' during the next time period is different for 'in' and 'out' animals

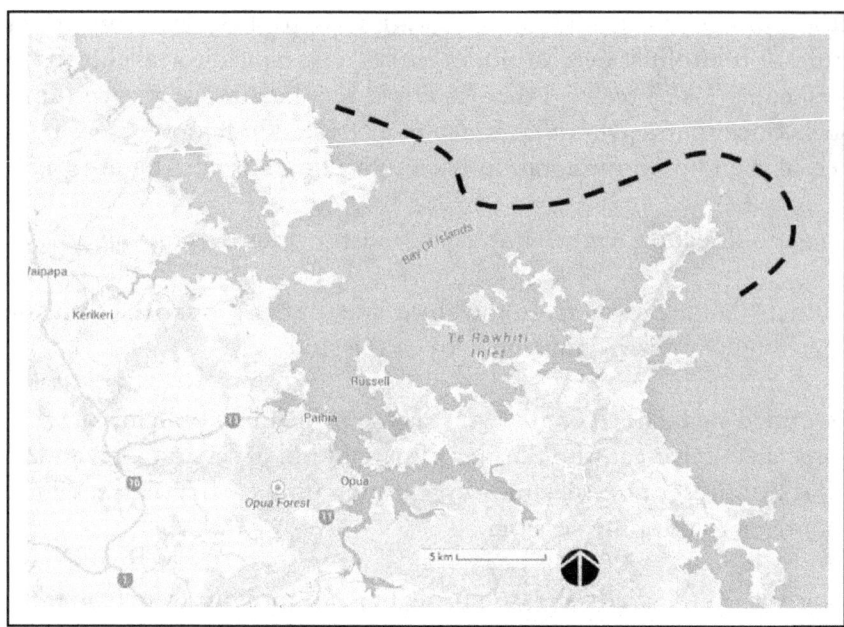

Figure 13.6: *The study site of Tezanos-Pinto et al. (2013) where dolphins were sampled in a bay from tourist vessels using photos. The dotted line shown represents a hypothetical extent of the sampling area, with the super-population extending beyond the sampling area.*

An example: bottlenose dolphins in a bay

Tezanos-Pinto et al. (2013) used the robust design to estimate population size of a coastal population of bottlenose dolphins (*Tursiops truncatus*) in northern New Zealand (Figure 13.6). The population was uniquely suited for the use of the robust design because:

- Super-populations of dolphins exists on the coasts of New Zealand. The populations do not emigrate permanently away from the coasts, and the various super-populations are known to be genetically isolated (animals do not move from one super-population to another).
- The specific super-population in the study was sampled through photos taken by tourism vessels and other vessels in the Bay of Islands. The super-population extends beyond the bay, so only a fraction of the super-population was available to be encountered during specific, short time intervals.
- Biologists had a goal of assessing population trends through time, which required estimates of N that were not biased by temporary emigration and immigration.

Figure 13.7: *Bottlenose dolphin (Tursiops truncatus) dorsal fins with distinctive natural markings. Copyrighted photo courtesy of Gabriela Tezanos-Pinto.*

The team of scientists scoured photos taken by scientists and by tourists in dolphin-watch vessels in the region. Dolphins were identified from photos by unique nicks and cuts on the trailing edge of the dorsal fin (Figure 13.7; Tezanos-Pinto et al. 2013). The biologists then used their estimates of gamma (γ) to estimate the size of the super-population (all coastal dolphins in the region; Figure 13.8).

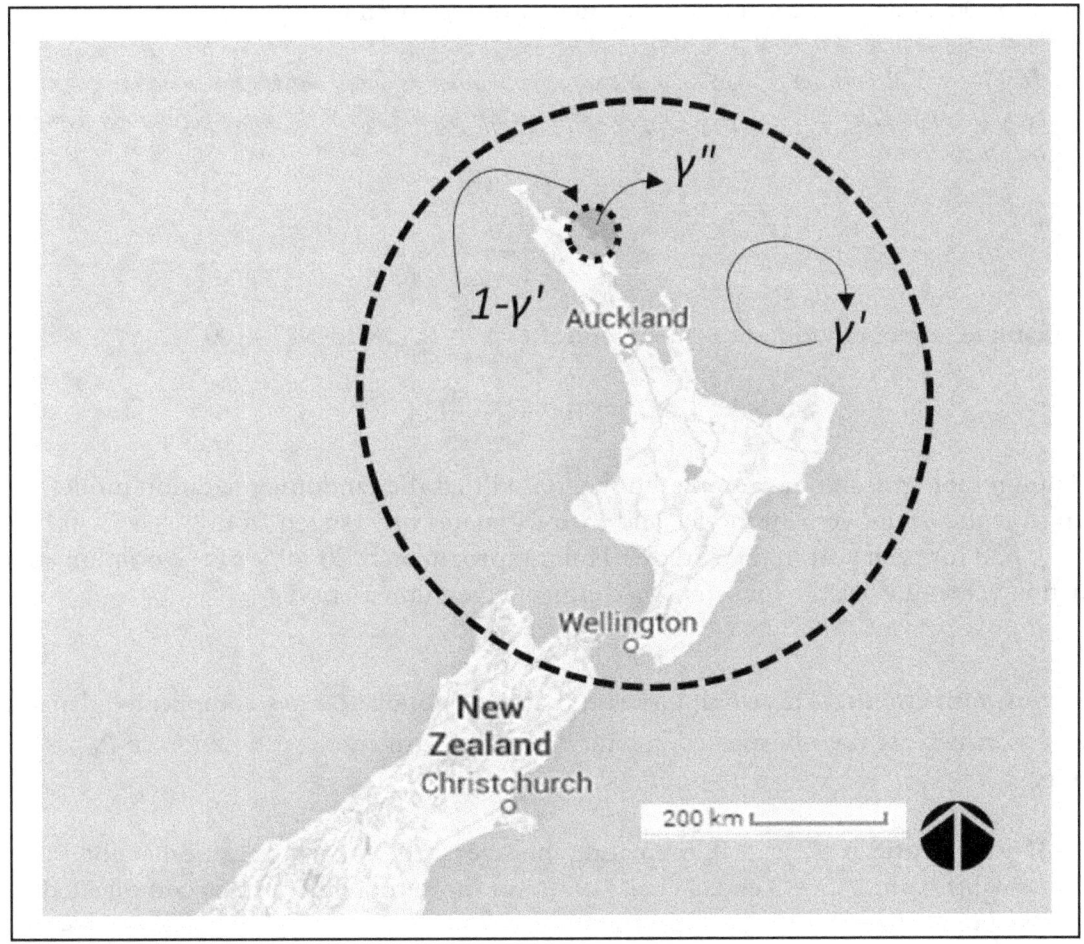

Figure 13.8: *The hypothetical borders of the study site (small circle) and super-population (large circle) of Tezanos-Pinto et al (2013) shown in the context of the robust design structure used in their analysis. Dolphins in the study area during a given time period could temporarily emigrate (γ'') before the next time period. Animals outside the study area in a given time period could return to the study area ($1 - \gamma'$) or remain outside of the study area (γ').*

Estimating N_{sp}

As a final comment, let us examine the three basic movement models to describe how we could estimate N_{sp}—the size of the super-population. Again, we start with the fact that our robust design analysis will provide us with estimates of N for each primary sampling period. We will also receive, from the most general model, estimates of γ for each time period.

No movement. This is the easiest result to interpret, but we might predict it does not occur very often, because animals move—the basic reason for using the robust design model!.

> *If $\gamma' = \gamma''=0$, there is no temporary emigration, and animals remain on the study site. If animals move, they remain available during all primary sampling periods—even if they may move outside the sampling area.*

$$N_{sp} = N$$

Random movement. The estimation of N_{sp} is fairly straightforward under this model, which makes this model desirable for the convenience of the research biologist.

> *If $\gamma' = \gamma''$: A constant proportion of the super-population is found outside the sampling zone between periods. Indeed, we can view γ as the proportion of the population of the super-population found outside the study area.*

$$N_{sp} = \frac{N}{1-\gamma}$$

As an example, under random movement with $\gamma' = \gamma'' = 0.13$ and $N(.)=100$

$$N_{sp} = 100/0.87 = 114$$

The dolphin biologists in our example, above, found that the random emigration model was chosen over the no-movement model, and their estimates of γ ranged from $\gamma' = \gamma'' = 0.182$ to $\gamma' = \gamma'' = 0.820$ for each year in their study. Thus, approximately 20-80% of the dolphin super-population was not available for sampling during a given time period.

Markovian movement. This result tells you that your population has complicated dynamics of movement, which are not unexpected in nature. But, unfortunately, estimation of N_{sp} is difficult.

> *If $\gamma' \neq \gamma''$ and if $\gamma', \gamma'' \neq 0$: estimating the size of the super-population is not straightforward. See Kendall et al. (1997) for more thoughts on this complicated dynamic.*

Robust design model assumptions

As a combination of closed and open methods, robust design shares **assumptions** with simpler models discussed in previous chapters. In addition:

- the population is assumed closed to additions and deletions across each secondary sampling session within a primary period
- survival rates are assumed to be the same for all animals in the population, regardless of availability for capture
- permanent emigration out of the super-population influences the survival estimate (incorporated as mortality)
- individual marks are required; thus, marks must not fall off and must be recognizable

We caution you to consider the number of parameters estimated by this model structure as you design your study. Tezanos-Pinto et al. (2013, dolphin example above) reported 188 parameters (including, for example, 28 estimates for N and 27 estimates for γ for specific time periods during the study) estimated for their top-ranked model. One of the lower-ranked models had 531 parameters!

How can you determine how many animals to tag? Adequate sampling necessitates the ability to make a good guess (at a minimum) as to the movement dynamics of your population—if a large number of animals are not available for capture during a given time period, you will need to tag more animals to obtain a useful sample of captured (and recaptured) animals! We encourage you to review the literature—look at the sample sizes, and the precision of estimates used by others with similar research designs.

Conclusion

The typical use for robust design methods is for monitoring or sampling schemes with annual, or long-term periodic sampling and short bursts of *effort* during the year. The method is useful for situations in which the sampling gear may not 'cover' the entire area of interest (e.g., the super-population's full extent). And, the method is used by biologists who are interested in closed and open parameters (movement, survival, N).

We note, in this last chapter dedicated to mark-recapture analyses of survival, that many model structures are available for survival estimation. The method selected by the biologist should depend on the design of the project in terms of **types of tags** (e.g., is known fate possible?), the **timing of sampling** (e.g., is robust design possible?), and the **design of sampling** (e.g., is multistate possible?).

We hope that you now have a basic understanding of these model structures. At this point, we also hope that you can see the creative 'sport' that exists as you think about the parameters of interest and how to get the information you need to describe the dynamics of your population!

References

Bailey, L. L., T. R. Simons, and K. H. Pollock. 2004. Estimating detection probability parameters for Plethodon salamanders using the robust capture-recapture design. Journal of Wildlife Management 68: 1-13.

Kendall, W. L., J. D. Nichols, and J. E. Hines. 1997. Estimating temporary emigration using capture-recapture data with Pollock's robust design. Ecology 78: 563-578.

Pollock, K. H. 1982. A capture-recapture design robust to unequal probability of capture. The Journal of Wildlife Management 46: 752-757.

Tezanos-Pinto, G., R. Constantine, L. Brooks, J. A. Jackson, F. Mourão, S. Wells, and C. Scott Baker. 2013. Decline in local abundance of bottlenose dolphins (*Tursiops truncatus*) in the Bay of Islands, New Zealand. Marine Mammal Science 29: E390-E410.

For more information on topics in this chapter

Conroy, M. J., and J. P. Carroll. 2009. Quantitative Conservation of Vertebrates. Wiley-Blackwell: Sussex, UK.

Kendall, W. 2014. Chapter 15: The 'robust design' *In* Program MARK: a gentle introduction, 12th edition, Cooch, E. and G. White, eds. Online: http://www.phidot.org/software/mark/docs/book/pdf/chap15.pdf

Williams, B. K., J. D. Nichols, and M. J. Conroy. 2002. Analysis and management of animal populations. Academic Press, San Diego.

Citing this primer

Powell, L. A., and G. A. Gale. 2015. Estimation of Parameters for Animal Populations: a primer for the rest of us. Caught Napping Publications: Lincoln, NE.

Ch. 14

Introduction to survey sampling[1]

"You can observe a lot by just watching"
-- Yogi Berra

Questions to ponder:
- *What is the difference between abundance and density*
- *How are a census, a sample survey, and an index similar and different?*
- *How do 'availability' and 'incomplete detection' affect estimates of abundance or density?*
- *What types of surveys correct for incomplete detection or availability?*
- *Do I need to correct for incomplete detection or availability?*

"Abundance" estimation: a background

One of the primary field techniques used in wildlife and fisheries research is some kind of survey that has the objective of either (1) counting a population or (2) estimating the density of the population or (3) obtaining some index to the relative population size. Biologists haul in nets of fish, walk forests and grasslands to count songbirds, use aerial photographs of seal colonies, conduct strip surveys for whales or dolphins, scuba along reefs to count moray eels (*Muraenidae*), listen at night on the edge of wetlands to frogs, collect unheard (by human ears) ultrasonic calls of bats, or conduct visual surveys of animals like black-tailed prairie dogs (*Cynomys ludovicianus*) or greater prairie-chickens (*Tympanuchus cupido*) on the breeding grounds. There is a method to try to count any species that exists, thus it is particularly important to carefully match the survey method with the biology of the species of interest.

We'll identify C as the number of animals we count (or hear) in a particular area, and N is the true abundance—the number of individuals in that area. We know that in most cases, our count (even though we try very hard) does not equal the number of animals in the population.

---

[1] *With thanks for content to David Smith, Max Post van der Burg, John Carroll, Michael Conroy, and Gary White*

A bit about terminology:

Census versus **abundance estimate**—a census is a full count of a population, which we assume to be complete. If our assumption is correct, there is no sampling or sampling error. So, $C = N$. In contrast, an abundance estimate is derived when a population is sampled by counting, so we assume $C \neq N$. But, we use the sample count to derive an estimate of N. We can think of the difference between C and N, loosely, as **detection error**.

Typically, a census is only possible for small areas and/or limited populations (e.g., an endangered species on a small island).

An **index** is obtained when a count is conducted, but detection rate is not measured. Because we don't know the detection rate, we don't know much about the exact relationship between C and N. But, we can say that $C \leq N$.

An example of an index would be the Rural Mail Carrier Surveys reported to Nebraska Game and Parks Commission each year by postal workers who are asked to note the number of ring-necked pheasants (*Phasianus colchicus*) they see. Since the 1950s, they have written down how many pheasants they see on their routes.

If we assume that the relationship between true pheasant abundance in Nebraska and this index remained the same over time, we would infer that the pheasant population has really decreased since the 1950s.

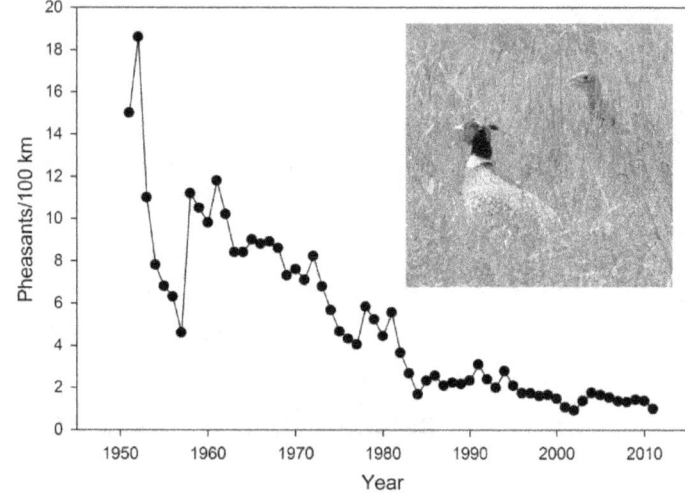

Figure 14.1: *The statewide index for ring-necked pheasants (Phasianus colchicus) in Nebraska, as determined by rural mail carrier surveys (number of pheasants per road distance traveled). Figure used with permission of Nebraska Game and Parks Commission. Photo of pheasant is a public domain image.*

But, indices have many possible relationships with true population density. Biologists usually assume a linear, positive relationship (A, Figure 14.2).

But, what if mail carriers became much more fastidious about looking for pheasants (their detection rates went up because they were hyper-vigilant; Figure 14.2B) as pheasant populations went down? Well, that would mean that their populations have actually decreased more than what the figure of pheasants/100 km shows. Or, what if mail carriers stopped looking for pheasants when populations were low—perhaps they felt that it wasn't worth their concentration and effort (Figure 14.2C). If detection rates were low at low population levels, the index shows a starker decline than really exists.

This is the problem with an index. Unless it is 'grounded' by some form of repeated-sampling to calibrate the index to give it rigor, these types of questions abound.

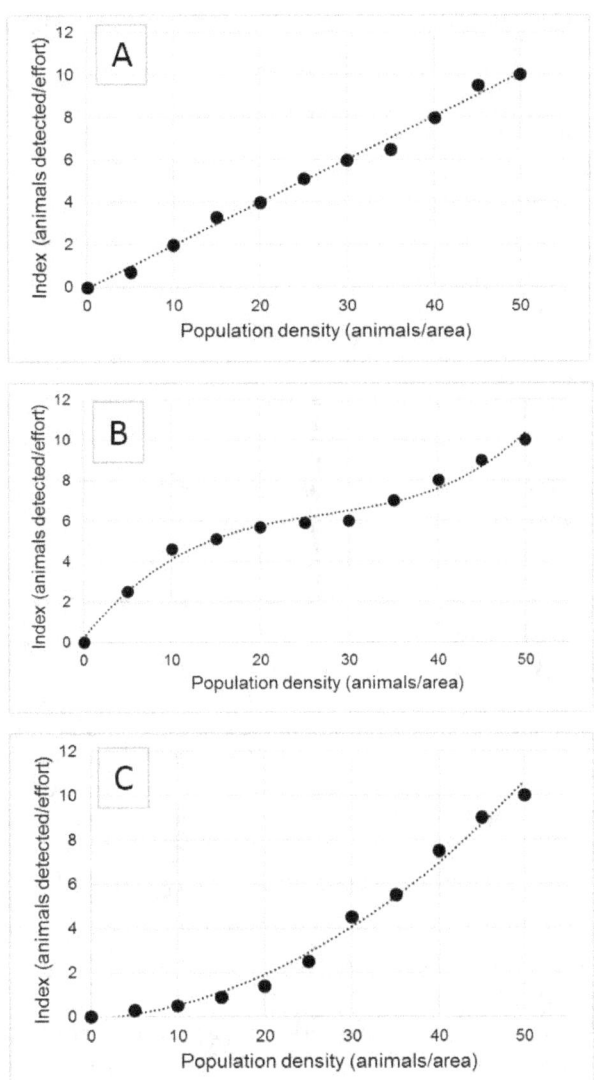

Figure 14.2: *Possible relationships between an index and the actual population density (after Conroy and Carroll 2009).*

Fisheries biologists think about indices a lot—perhaps more than wildlife biologists. **Catch-per-unit-effort (CPUE)** is a grand example of an index. CPUE is a measure of how many fish were caught per net-night (one net in a lake or stream for one night equals one **net-night**). But, most fisheries biologists are quite aware that many things can cause catch-per-unit-effort to change through time…even if abundance of the fish population stays the same.

Just as with bird survey indices, CPUE and abundance do not always have a linear relationship (Harley et al. 2001). Imagine a lake with a large population of fish. Sixteen nets are set out, and a CPUE is established for that population. Now, a disease goes through the population of fish and the population dramatically decreases in size (abundance declines, Figure 14.3). Those same

16 nets are put out again. If the fish move at random around the lake (like you might want to assume statistically), the CPUE might also decrease linearly. But, if those fish are spatially relegated to an area of the lake where you don't have your nets, your CPUE is going to decline dramatically...it could be zero when the population is not zero (*hyperdepletion* curve, Figure 14.3).

Alternatively, you might think of a situation where fish might move more if their competitors have been removed from the system. If abundance were to decline in the lake, the movement of the fish might keep the CPUE at almost the same level (the *hyperstability* curve in Figure 14.3), and a fisheries manager might not realize abundance had dropped. In fact, Harley et al. (2001) found evidence that the CPUE-abundance relationship is different for different species of fish in the same sampling scheme.

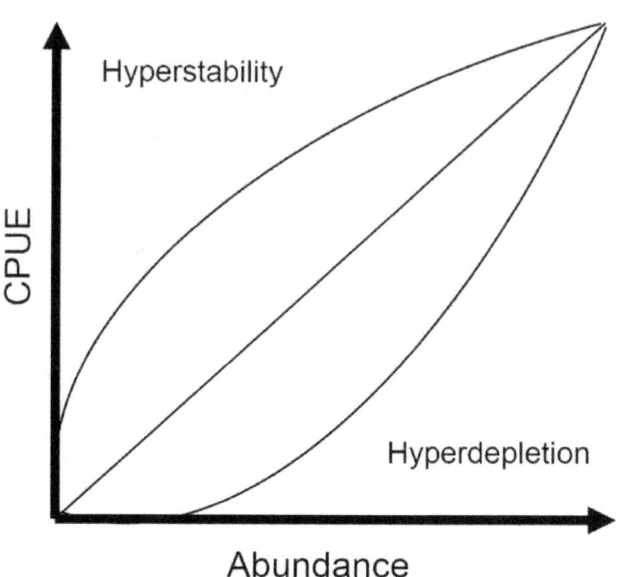

Figure 14.3: *Possible relationships between catch-per-unit-effort indices and actual fish abundance (after Harley et al. 2001). Hyperstability occurs when catch per unit effort remains stable despite significant declines in fish abundance. Hyperdepletion is when CPUE declines faster than abundance and using these data in an assessment can produce overly pessimistic estimates of abundance.*

So, indices can be problematic to use and interpret in some cases. But, lest we throw out indices in favor of more complicated methods, one should read Johnson (2010), in which the author effectively argues "*....although criticism of the thoughtless use of indices is welcome, their wholesale rejection is not.*" Indeed, there is a level of intellectual snobbery from the quantitative elite (a club that you are on the path to membership, now!) towards indices. Your authors have been guilty of such thoughts in their careers as well. However, logistical and cost issues can sometimes prevent the use of more complicated methods. And, it is not hard to argue that effective management decisions have been made, worldwide, using indices as the information to feed into decision processes. The fact is that indices are typically cheaper to conduct, and they are used commonly in situations where funding is limited.

Johnson's (2010) main point is that biologists often reject indices for their uncertain relations to true abundance in favor of estimation processes, but these estimation processes may also be fraught with their own issues (e.g., assumptions of closure for *N*-mixture models or 100% detection on the transect line or accurate distance estimation in distance methods) as you will see in future chapters. So, Johnson (2010) has a point: an index would be no worse than a poorly executed survey (i.e., one that grossly violates a method's assumptions) used to estimate density, for example. Therefore, we should learn how to plan, conduct and analyze surveys (of any type) properly.

Abundance estimation: from incomplete samples

We have postulated that our counts do not equal the population size in most situations ($C \neq N$). We can formalize this a bit more as:

$$\hat{N} = \frac{C}{P_{area}\hat{P}_d}$$

Here, we correct the count (C, number of animals counted) with two proportions, P_{area} and P_d.

First, we can think of our sampling design. If we are sampling less than the entire area of interest, then we need to adjust our count by that proportion, P_{area}. So, if the population (N) was known to cover 100 hectares, and we only sampled 50 hectares, we would logically assume that we would count half of the animals (assuming they are spread randomly throughout the entire area, and assuming we have perfect detection of animals in our sample area). This correction is called **design-based sampling**, because we can get this correction factor easily by looking at the design of our sampling (e.g., a map).

Now, what about **detection**, P_d? What if our detection is not perfect? What if our capabilities only allow us to see 25% of the animals in our sample site? We need to correct for this, obviously. So, we need to inflate our count by an additional 75% to make up for the missing animals on the study site. This factor is not easily gotten, however. We are not privy to our personal detection rate! So, we have to estimate it from patterns in our data, in some manner. So, we call this correction factor **model-based sampling**.

Design-based sampling

Many game counts in southern Africa are conducted by counting game in visible strips (the method is called **strip counts**). The width of the strips depends on how far a person can see in various habitats. This method can be used along roads, fairly easily. Obviously, you might end up 'pushing' animals ahead and some may remain unseen (so the count may be biased low). And, small game animals may have lower detection rates than larger game. But, if we assume that we see all the game in our strip, you simply 'correct' your count by extrapolating your count across the remainder of the study area.

$$N = nH/h$$

Where we define the following statistics:

N = Number of game of each type on the area
n = number of each type of game per strip
H = Surface area of farm/pasture/reserve
h = surface area of sampled strip

Fisheries biologists often use a similar design-based extrapolation when they sample a portion of a lake and extrapolate the count to the entire lake—just as the count in the strip is extrapolated to the pasture, above. We emphasize that this basic design-based sampling correction does not correct for animals missed because of incomplete detection.

You try it!

Try your hand at a design-based correction to the following survey data:

> Total area of pasture: 2110 ha
> Total area of strips sampled: 791 ha
>
> Survey data: we saw 88 impala (*Aepyceros melampus*) and 22 greater kudu (*Tragelaphus strepsiceros*)

How many impala and kudu are in the entire pasture? The answer is at the end of this chapter.

Abundance estimation: generalized

Of course, life is always more complicated than we would like it to be. In fact, the **detection probability**, P_d, in the previous section really refers to three distinct probabilities:

$$\hat{N} = \frac{C}{P_{area}\hat{P}_d} \qquad \hat{N} = \frac{C}{P_{area}\hat{P}_{detection}\hat{P}_{availability}\hat{P}_{presence}}$$

Presence probability: the probability that an individual which uses the sample site during some portion of its life is at the sample site during a time of interest (such as when you survey it).

Let us say we sample a portion of an animal's home range, but it is at the other edge of its home range during the sampling period (Figure 14.4). Theoretically, it should be enumerated as a member of the population, but we miss it because of this spatial problem.

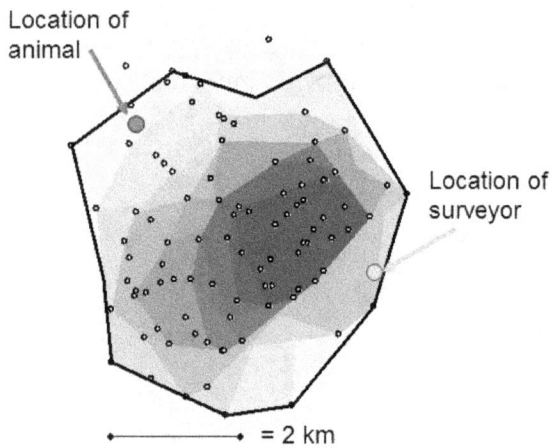

Figure 14.4: *Depiction of concept of presence probability. Animal is not always near the survey point, but uses the survey point.*

Current statistical analyses of survey data do not adjust for the portion of animals that use a given site (at some point in the time period) that are not present when you survey. *This problem actually represents a mismatch between the scale of the animal's movements and the survey instrument.* Thus, the best way to eliminate this problem (goal: have the probability of presence be 100%) is to use a larger scale of sampling (e.g., transects instead of point counts) or to adjust the length of time that you survey (so an animal can move back to our sampling point). Alternatively, you might consider using a robust design approach (see Chapter 13) to account for the temporary emigration from your survey area.

Availability probability: the probability that an animal at your study site is available to be sampled.

An easy example of availability bias in a survey is to think about any animal that lives in a hole in a tree or in the ground. In the western United States, such a species is the black-tailed prairie dog (*Cynomys ludovicianus*; Figure 14.5).

At any moment in time, even if you detect the animals above ground perfectly, a portion of the population is below ground. They are not available to be counted. And, that portion is obviously very important. Is 90% of the population above ground? Is it only 10% above ground? That's a big difference, and would affect our estimate of the population size.

Figure 14.5: *Black-tailed prairie dog, photo by Ron Singer, US Fish and Wildlife Service (public domain image).*

Detection probability: the probability of seeing/observing/hearing an individual, given that it is present and available at the sample point.

We can reference our prairie dogs again. Imagine that you are standing at the edge of a large colony (prairie dogs live in colonies of multiple burrows) and you are looking out across the colony. Prairie dogs are smallish creatures and they are brown—they blend in with the ground easily. As the distance from you to any prairie dog increases, odds are that you'll have a poorer chance of detecting the critter.

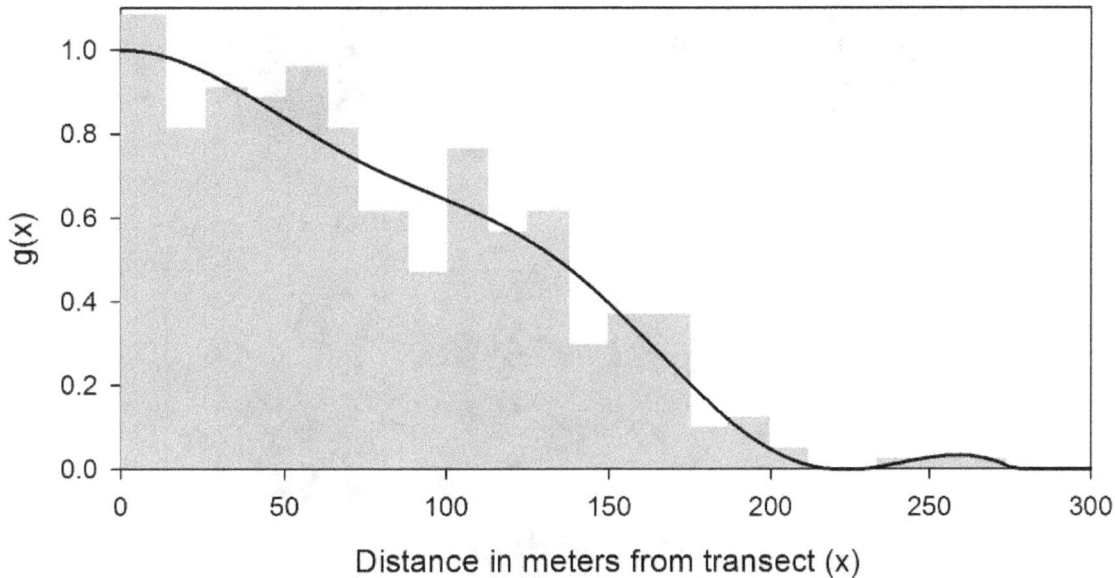

Figure 14.6: *Count data (gray bars) and detection function, g(x) for western meadowlarks shown as distance from a transect line in a grassland habitat increases (Kempema 2007).*

The count data in Figure 14.6 (gray bars) are for a loud-singing, yellow-breasted grassland bird, the western meadowlark (*Sturnella neglecta*). These data are from line transects through grasslands in Nebraska, USA (Kempema 2007). Unless meadowlarks were hanging out near the transect line in higher numbers than in other spaces in the pastures (very unlikely!), this figure shows a stark decrease in detectability as distance from the line increases. In fact, the problems with distance occur even for big animals…like kangaroos, deer, and whales (Buckland et al. 2001).

Here is a list of <u>other</u> potential reasons that detectability may not be 100% on a survey:

- Secrecy/crypsis – cannot see/hear animals easily
- Observer ability – lack of training or skills prevent detection
- Visual obstructions – grass/trees get in the way
- Wind – animals are harder to hear when windy, and less active so harder to see
- Weather – animals less/more active because of temperature or precipitation
- Time of year – breeding vs. brooding affects activity or color of animal

Detection can be a fairly complicated dynamic in the real world. As an example, Rigby (2014; USFWS teleconference transcript, used with permission) provided some very unique depictions of the spatial variation of probability of detection of a grassland bird around a point count during conditions of high and no wind (Figure 14.7). At higher winds and with the observer upwind from the bird, both dickcissel (*Spiza americana*) and eastern meadowlark (*Sturnella magna*) had very low probabilities of detection. Such dynamics can be accounted for with analyses and/or accounted for with survey design or protocols (e.g., no surveys conducted during high wind speeds).

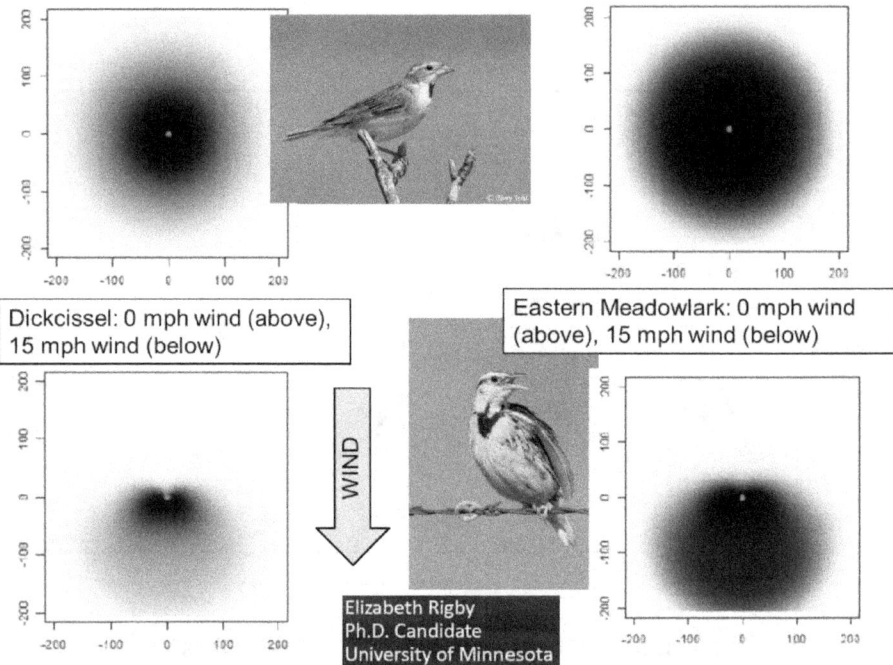

Figure 14.7: *Spatial depiction of model predictions of variable detectability (darker shows higher detection probability) for two grassland songbirds, eastern meadowlark and dickcissel, under no-wind and 15-miles per hour (~24-km per hour) wind conditions. Wind direction is shown by arrow. At high wind speeds, the birds were virtually undetectable when the observer was upwind. From Rigby (2014), used with permission.*

How to select a survey method

One should always have a goal. So, what is the goal for your survey? Do you want to estimate population size? Should you consider a closed mark-recapture approach? Or, is a survey better?

Is density (number of individuals per unit area) your interest? We should note that density is often hard to estimate. For example, if you use a trap grid to estimate abundance (N) of small rodents using mark-recapture, how can we find the density? Well, we could approximate the size of the area trapped, right? Well, what is that area? What is the 'effective' size of the grid? How far did rodents come to be trapped? See the problem? It is hard to determine the appropriate area to use to calculate density.

Abundance or relative abundance is often easier to estimate. But, density allows direct comparisons among studies.

Effective sampling begins with clearly defined study objectives and a good understanding of the species being sampled (biology, natural history, etc.). Some survey or research questions might be as follows:

- What is the current density of an endangered species in an area of potential impact?
- What is the recent trend in abundance of a species in an area of interest?
- Why is abundance changing?

We might be doing a survey because we have management needs to make decisions based on monitoring. Perhaps we need to provide information on the current state of a system. Will we be evaluating management performance based on changes in population size or density?

All of these questions are important as we plan for the type of survey we need—a census, an index, or an estimation method.

Simple surveys: assuming 100% detection

Quadrat sampling represents the simplest type of sampling—identify some areas that can be defined (squares in Figure 14.8). Then, the number of organisms in the quadrats is counted (a complete count, or local census, within the sample area). This is a spatial sampling approach, and the true abundance can be estimated by extrapolating the abundance (n) from the quadrats (as in our southern Africa example). Plant biologists use this method often, and they assume that detection is not a problem.

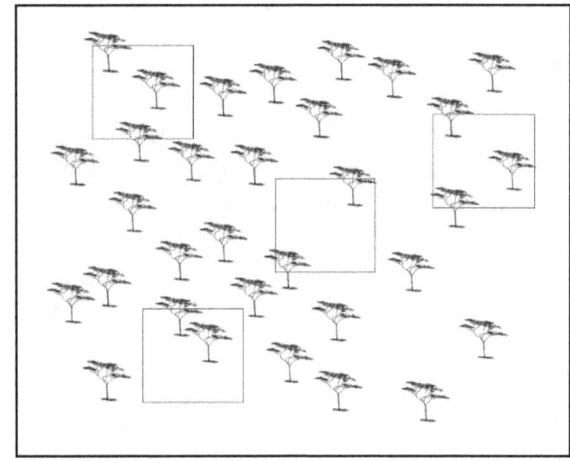

Figure 14.8: *A population of trees sampled with four quadrats (squares).*

Density, in a quadrat survey, can be quickly calculated as:

$$\hat{D} = \frac{n}{area}$$

The next step in surveys was to extend the quadrat in space along a route that was walked or traversed in some other way (Figure 14.9). This survey is called a **fixed-width transect census**. Animals are counted from the transect (of length: L) to a certain width, w. For example, an avian biologist can walk through the woods and count woodpeckers within 20 meters of a trail. Or, a fisheries biologist can electroshock along a transect of a width covered by the electrical current from the boat. The fish are collected and counted. The width of the strip is typically chosen to attempt to assure 100% detectability within the strip. Density is estimated as:

$$\hat{D} = \frac{n}{2Lw}$$

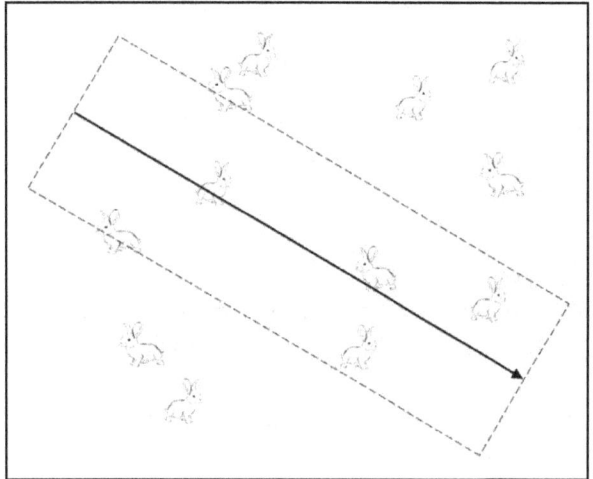
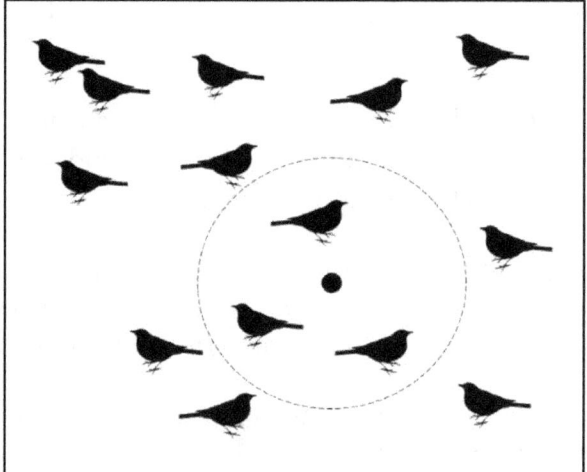

Figure 14.9: *A population of animals is sampled as a complete count (census) along a transect with a fixed width and known length.*

Figure 14.10: *A population of animals is sampled with a complete count (census) from a point location using a sample circle of fixed radius.*

A similar method to the fixed-width transect approach is the **fixed-radius point count**. These are used often by avian biologists. A radius, r, is chosen, to which the biologist feels comfortable that they can see/count all birds (Figure 14.10). Density is calculated as:

$$\hat{D} = \frac{n}{\pi r^2}$$

Imperfect detectability

Figure 14.11: *Comparison of visibility during a survey of birds on a coastal beach in Washington, USA, during normal (left) and foggy (right) conditions. Photos by L. Powell.*

We've already established that detectability can be an issue. It is usually easy to assume that some factor will prevent the surveyor from detecting all the animals in an area. In fact, sometimes detectability changes during a survey. For example, the photos in Figure 14.11 were taken during a survey of peregrine falcons (*Falco peregrinus*) and bald eagles (*Haliaeetus leucocephalus*) along the beaches of coastal Washington. Fog rolled in during the survey and dramatically changed the distance at which birds could be seen from the car. And, as you might expect, the surveyors counted fewer birds when they were in the fog.

So, what kind of survey methods can help when detectability is an issue? You have several choices, and each has its advantages and constraints.

The **known component method** is one of the simplest methods. It requires the use of a 'known component' that can be used to correct the count for the detection problem.

This is another method used in southern Africa. Let us suppose that 114 plains zebra (*Equus quagga*) were introduced to a farm during the week before a survey was conducted (i.e., the "known component"). During the road strip survey, 78 zebra were counted along with 283 Cape buffalo (*Syncerus caffer*).

For this method, we can establish the following statistics:

N_x : Unknown population size of species x
N_y : Population size, known, of species y
n_x : Number counted of species x
n_y : Number counted of species y

If we assume that both species have <u>equal</u> detection rate, we can use the fact that we counted 78 of the 114 zebras (68%) to help estimate the number of buffalo. This method would not work well for predator (low detection) vs. herbivore (higher detection). And, of course, the requirement is that you <u>know</u> the number of one of the species being surveyed, which is rarely the case.

But, if the situation allows, we can estimate:

$$N_x = \frac{(N_y)(n_x)}{n_y}$$

A version of the known component method has been used to correct aerial surveys of waterfowl in North America. You can imagine flying over a wetland in an airplane, trying to count the number of ducks. By putting a known number of duck decoys (N_y) out in a certain area (the number of decoys not known to the person counting from the airplane of course!), the known component method can be used to estimate the population size of real ducks in the area (N_x) if we assume the counts of ducks (n_x) and decoys (n_y) are susceptible to the same probability of detection.

Distance sampling was developed to account for the declining detectability with increasing distances away from a transect line, as shown with the meadowlark example earlier in this chapter.

Using distance sampling, a surveyor walks a transect line, as in the transect census method. But, in this method we do not assume we count all of the animals. In fact, we assume we will miss some individuals, as seen in Chapter 19 with elephants. The data collected are simply the number of animals seen, and (for each animal seen) the perpendicular distance from the transect to the animal (x). The pattern of observations allows us to estimate a detection probability, so we can adjust our count to reflect the density of animals near our transect line. This method can also be used with point counts, as we will demonstrate in Chapter 19.

The **double-observer method** uses two observers (Figure 14.12), and the analysis uses the pattern of imperfect observations from both observers to estimate the detection probability. We assume that both observers may not see all of the animals. Both observers record 'seen' animals, and they then reconcile which animals both observers saw. As we will discuss in Chapter 16, the analysis is based on mark-recapture theory—you might think of each animal observed as having a 'capture history' where the 1's and 0's show whether observer #1 (first column of capture history) saw the animal and whether observer #2 saw the animal (second column of capture history). For example, a capture history of "11" indicates that both observers saw the individual.

Figure 14.12: *Depiction of potential results of a double-observer survey in which animals can be sighted by either or both observers.*

Double-observer survey possible results	Represented as capture history	Number of individuals with capture history (*Figure 14.12*)
#1 yes, #2 no	10	1 individual *(dark arrow)*
#1 no, #2 yes	01	2 individuals *(gray arrows)*
#1 yes, #2 yes	11	2 individuals *(dashed lines)*

Fisheries and small mammal biologists have been using **removal methods** (*"Zippin depletion method"*) for decades. The method uses repeated samples, one after another during a very short time span, and might result in capturing all the animals in the area if sampling were to continue for long enough. Animals are "removed" (not replaced) after sampling.

Avian biologists have extended the removal sampling method to bird surveys, by using 3-4 time intervals during the point-count survey as individual sampling occasions. The logic is that you will begin to see fewer and fewer 'new' birds as the survey goes through time (point-counts are often 5 or 10 minutes long). Birds not initially 'available' (hiding in the grass) will eventually pop up. So, this method accounts for both availability and detection, as we will discuss in Chapter 17.

A typical set of "removal" data might look like this:
- Time interval 1: 10 new individuals
- Time interval 2: 5 new individuals
- Time interval 3: 4 new individuals
- Time interval 4: 1 new individual

***N*-mixture methods**: this method also estimates abundance by using multiple visits to the same points and multiple samples at the study area (envision a grid on which each point is sampled multiple times). By visiting within a short time period (so that we can assume 'closure' of the population), we can develop a count history with the site samples as rows and the visits as columns:

```
              Visits
    Site A  0  1  2  0
    Site B  0  0  0  0
    Site C  0  2  0  0
```

Here, we know (for Site A, the top row) that the minimum number of animals on-site is 2. We also know that we missed at least one animal during the previous visit (we only saw 1), and we missed all (at least 2) animals during the first and last visits. With this information, a corrected count is developed. We will discuss this method in Chapter 18.

Counting too many animals?

To this point, we have assumed that our counts are equal or less than the number of animals in the population. And, this is almost always true. However, there is one type of survey for which the count is usually <u>greater</u> than the number of animals ($C>N$): surveys of waterholes.

Here's how a waterhole survey typically works: biologists sit in a blind (also called a "hide" if you are in southern Africa) and they count all animals seen in 48 hours (Figure 14.13). The time period of 48 hours is chosen because all animals are assumed to come to drink at least once during 48 hours. Sampling occurs during the full moon for night viewing and counting.

Figure 14.13: *The view of a waterhole at Waterberg National Park in Namibia during the dry season. Photo taken by L. Powell from the blind. Salt blocks are scattered between the waterhole and the blind to encourage rhinoceros to come closer at night for better individual identification.*

Of course, some animals come to drink more than once during the 48 hours, so you end up with more animals counted than are in the population. So, you must use the drinking frequency (H, hours between drinks) of each species to correct the count (relative to the number of hours, $H_{sampled}$, during which you counted).

$$N = C \bullet \frac{H_{betweendrinks}}{H_{sampled}}$$

For example, if you count 50 giraffe (*Giraffa camelopardalis*) in 48 hours, and giraffes drink once every 24 hours, then there are 25 giraffes in the population: 50 x (24/48) = 25

Similarly, if you count 15 rhino in 48 hours and rhino drink once every 48 hours, then there are 15 rhino in the population: 15 x (48/48) = 15

Conclusion

We will cover, in more detail, many of these model-based survey methods in the following chapters. There is not a "best" survey method. The method used should be based on your objectives and the natural history of your study species. And, each survey process will have constraints. For example, if you are the only person available to do surveys, you cannot use the double-observer method. Below is a comparison of the main methods we have introduced in this chapter.

Method	Accounts for incomplete detectability	Accounts for incomplete availability	Requires >1 person	Requires >1 visit	Simple data recording
Strip census					X
Distance	X	X[a]			
Double-sample	X		X		X
Double-observer	X		X		
Removal	X	X			
N-mixture	X	X[b]		X	X

[a] Advanced form of distance analysis with removal sampling accounts for availability; typical distance sampling does not.

[b] Advanced, multi-season form of N-mixture model estimates availability; single season N-mixture model does not.

Answers: You Try It!

Total area of pasture: 2110 ha
Total area of strips sampled: 791 ha
Survey data: we saw 88 impala and 22 kudu.

How many impala and kudu are in the entire pasture? First, we come up with the multiplier to use to extrapolate our counts to obtain the population size: 2110 ha ÷ 791 ha = 2.67. That is, our pasture is 2.67 times as large as our sample area. So, we multiply species' counts by 2.67.
  2.67 x 88 impala = 235 impala
  2.67 x 22 kudu   = 59 kudu

References

Buckland, S. T., et al. 2001. Introduction to Distance Sampling: estimating abundance of biological populations. Oxford University Press.

Conroy, M. J., and J. P. Carroll. 2009. Quantitative Conservation of Vertebrates. Wiley-Blackwell: Sussex, UK.

Harley, S. J., R. A. Myers, and A. Dunn. 2001. Is catch-per-unit-effort proportional to abundance? Canadian Journal of Fisheries and Aquatic Sciences 58: 1760-1772.

Johnson, D. H. 2008. In defense of indices: the case of bird surveys. Journal of Wildlife Management 72: 857-868.

Kempema, S. L. F. 2007. The influence of grazing systems on grassland bird density, productivity, and species richness on private rangeland in the Nebraska Sandhills. MS Thesis, University of Nebraska-Lincoln, Lincoln, NE.

For more information on topics in this chapter

Bart, J., S. Droege, P. Geissler, B. Peterjohn, and C. J. Ralph. 2004. Density estimation in wildlife surveys. Wildlife Society Bulletin 32: 1242-1247.

Thompson, W. L., G. C. White, and C. Gowan. 1998. Monitoring vertebrate populations. Academic Press, San Diego.

Williams, B. K., J. D. Nichols, and M. J. Conroy. 2002. Analysis and management of animal populations. Academic Press, San Diego.

Citing this primer

Powell, L. A., and G. A. Gale. 2015. Estimation of Parameters for Animal Populations: a primer for the rest of us. Caught Napping Publications: Lincoln, NE.

Ch. 15

Occupancy modeling[1]

"Presence is more than just being there."
-- Malcolm Forbes

Questions to ponder:
- *What is the definition of occupancy?*
- *How do I gather data for occupancy analyses?*
- *How important is the closure assumption for occupancy models?*

What is occupancy?

Occupancy modeling is a form of analysis that is unique among the chapters in this book. Rather than a focus on recording fates of individual animals in a cohort (mark-recapture) or estimating abundance or density (other survey methods), **occupancy data is recorded at the species level**. We define two parameters, one of which (ψ) has not previously been used in this manner in this primer (but see use of ψ in Chapter 12 on multi-state models):

ψ (psi): proportion of area, or patches, that is occupied. Strictly, when applied to a given sample site, **occupancy is the probability that <u>a site</u> is occupied by a species**.

p: probability of detection of the species, given that a species is present. We note that because of timing of surveys, p is actually a product of detection (the species is there and you see/hear it) and availability (spatially or physically present to be sampled).

Why estimate occupancy?

Ecologists have many reasons to be interested in the concept of occupancy, or probability of occurrence. First, for rare and elusive species that are difficult to sample, we often obtain 0's during surveys. Thus, it is hard to estimate abundance or density. And, because of the rarity, our interest in the species may be focused on "where is it" and perhaps environmental factors associated with its presence rather than "how many are there?"

From a monitoring perspective, we might use occupancy-type data for studies of geographic range and distribution. Over time, is the species' range shrinking or growing (is occupancy

---

[1] *With thanks for content to Larissa Bailey, Max Post van der Burg, Marc Kéry, and Therese Donovan*

declining or increasing?)? Thus, for example, we might apply occupancy models to study invasive species, spread of disease, or the effects of climate change on distribution of a species.

In addition, biologists might use occupancy analyses, with appropriate covariates to describe how occupancy changes as measures of habitat characteristics change. Is this grassland bird more likely to be found in areas with more bare soil? Thus, we can answer, broadly, questions of resource selection.

Last, we might focus on the estimates of p to make inferences about the species' behavior. For example, Finley et al. (2005) reported the detection probabilities of swift fox peaked in February and October during periods of dispersal. Detection was lower in June and July when swift fox were more likely to be in their dens. The pattern of detection was useful, above and beyond the information provided about occupancy.

A biologist new to monitoring might ask, **"If I want to determine if the species is at this location, why not just visit once to look for the species—and be done with it?"**

Of course, if you've been reading various chapters in this book, you can answer that question easily (see Chapter 14). We must account for **false negatives** (failure to locate a species that is actually present). And of course, we account for false negatives with estimates of **detection probability**. With proper design of a study, we should increase our confidence that a species is truly absent if we fail to document it.

In fact, false negatives are a big problem when working with rare and elusive species that may have very low detection rates. If we fail to account for detection, our data may bias habitat-use studies, and we may erroneously interpret naïve (inferred only from our raw data) site occupancy observations.

A binomial landscape

In the simplest description, occupancy modeling attempts to predict the actual presence or absence of a species with repeated presence/absence observational data. We can define the **latent state** as the true state of existence that is hidden or concealed from the research biologist. So, if we had omniscient powers, we would know the truth about the species occupancy patterns. But, we don't have x-ray vision or complete truth—we have to sample and learn from our samples. As we sample, we start to see patterns that can help us determine probabilities of occupancy. These patterns can be summarized in the form of **detection histories** for each of our sample locations—very similar to capture histories used in mark-recapture studies (see Chapter 7).

Consider the following data for 4 points at a study location that is sampled during 3 occasions:

Point	Latent State	Data (Detection History)
1	1	0-0-0
2	0	0-0-0
3	1	0-1-0
4	1	1-1-1

Our **naïve occupancy**, the proportion of points at which we documented the species (looking at our data), would be 50% for this small sample. Of course, we know that true occupancy (the latent state) is actually 75%. At point #1, the species was present but never seen (a false negative). At point #2, the species was not present, and therefore was never seen. At point #3, the species was present, but only seen once. And, at point #4, the species was present and seen all 3 times.

If this was our complete set of data, and if we had perfect knowledge of the status of our species' occupancy, we would infer that our detection probability is certainly not 100%. We missed the species completely at point #1. At point #3, we know the species is present, but we only documented it during 1 of 3 visits. Even without perfect knowledge of the latent state, we already know from point #3's observations that we do not have perfect detection.

Maximum "Eye-klihood," revisited

In Chapter 10, we discussed how it is possible to look at a set of capture histories and infer some basic qualities about your species of interest. Instead of waiting for the maximum likelihood estimate to magically appear from our software, we can use Maximum "Eye-klihood" to visually (with your eye—get the pun?) assess our summarized data, so let us conduct a similar thought-experiment here. Consider the following two sets of data from a set of 10 sample locations that are visited 3 times:

	Study 1	Study 2
1	111	010
2	011	001
3	111	101
4	110	100
5	111	001
6	101	010
7	111	100
8	111	001
9	000	000
10	000	000

Which of the above sets of data would you predict to have the highest encounter probability, p? Of course, it appears that Study 1 has higher encounter probabilities—when an animal is seen at a point, it is often seen at least 2 and usually 3 times. In Study 2, we see that most of our observations show the species documented only 1 of the 3 site visits. Thus, we know the species is present, but there is a lower chance of observing it.

More importantly, we see that both studies failed to document the species of interest at the 9th and 10th sampling location. Was it truly absent, or did we just miss it? **If you take nothing else from this chapter—the idea of occupancy modeling can be summed up in the following question:** *for which of the two studies are you more certain that the species is <u>actually absent</u> at the 9th and 10th sampling locations?*

The answer to the question should be fairly obvious—we can predict the species is *most likely* to be absent from sample points 9 and 10 for Study 1. Why? Because Study 1 has such high detection rates—the patterns from the other points tell us that if the species is present, we are very likely to see it. We didn't see it at points 9 and 10—and inferences from our other data suggest that the species may not actually be there. For Study 2, we have more uncertainty about points 9 and 10. We have such low detection probabilities at points 1-8 that it is actually fairly likely that the species may be present at 9 and 10—but we failed to see it.

We can see that both studies have 80% naïve occupancy, based on our raw observations—the species was observed at 80% of the sample locations for both studies. However, the pattern of observations from multiple visits suggests that Study 1 has a true occupancy of 80%, while Study 2 might have a true occupancy of more than 80% (perhaps 90% or 100%) because we were more likely to miss an existing member of the species in that study.

And, that is the basic idea of occupancy modeling—we use patterns of observations from repeated visits to gain information about our ability to detect a species when it is present. And, that information allows us to estimate how many of the sites with no observations might actually have the species present.

Basic sampling design

In the simplest form of research design for occupancy, known as "single season, single species", the biologist selects a random sample of sites within the landscape. The research team makes multiple visits to each site, and during each visit the surveyors determine if the species was detected or undetected. At least two visits are required, but more than 2 visits is likely to increase the precision of occupancy and detection probability estimates. With the data collected from the repeated visits, we can create an encounter history, for example: 00110 (species seen during the 3rd and 4th visits, but not the 1st, 2nd, or 5th).

It is important to note that after the species of interest has been detected, **sites must be resurveyed so detection probability can be estimated**. Prior to the advent of occupancy modeling, it was fairly common protocol to stop sampling a site once the species was observed---why continue to visit, because the species has been documented. True, if we are only interested in naïve occupancy, but **we need the 'absence' information for occupancy modeling**. So, do not end sampling after detection! (If the cost of making multiple trips is a big issue, it is sometimes possible to stop sampling after a detection is documented, *provided* there is sufficient 'absence' information in the overall dataset, see MacKenzie et al. 2006 for details on how to implement occupancy surveys).

Typically, all sites are sampled at each time period, although many software packages can deal with missing values (for example if sites are added after the first two time periods). Occupancy sampling can be direct (visual observation of the animal by sight or camera) or indirect (identification of tracks or sign or vocalization through recordings). The ability to sample the presence/absence of a species through indirect or remote camera methods has opened the field of occupancy to address questions that traditional surveys were unable to address, and can lead in exciting directions if studies are designed thoughtfully.

Notation

Occupancy models and likelihood statements will usually involve the following parameters:

s = total number of sites surveyed
s_D = number of sites where the species was detected at least once
x = number of sites occupied
k = number of surveys
ψ_i = probability that a species is present at site i
p_{ij} = probability that a species is detected at site i at survey j (given it is present)

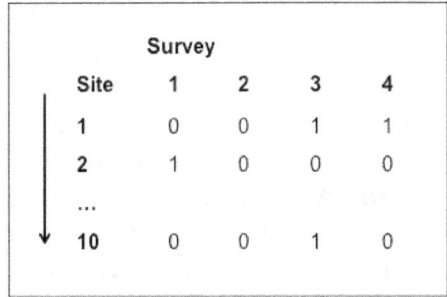

Figure 15.1: *Example design for an occupancy study with 10 sample sites and 4 visits to each site over time.*

Assumptions

Occupancy models carry the following **assumptions**:

- Sites are <u>closed</u> to <u>changes in state of occupancy</u> during sampling seasons
- Occupancy is constant across sites, unless modeled with covariates
- Detection probability is constant across sites, unless modeled with covariates
- Observers never falsely detect a species when it is absent
- Surveys and sample sites are independent

We note that **the closure assumption for occupancy models must be evaluated at the species level**—we are not concerned with individuals that might leave the area. Instead, we assume a species is either present or absent at a site during all sampling periods.

Options are typically available for analysis of occupancy data as **single-season** (data collected in one closed sampling 'season') or **multiple-season** (data collected in multiple closed seasons). The closure assumptions for a season still apply within each season for samples collected in multiple seasons, analogous to the robust design of Chapter 13 (MacKenzie et al. 2003).

The closure assumption is most lenient for occupancy modeling, relative to most other closed-type surveys or mark-recapture models, because closure is assumed at the species level rather than the individual level. Obviously, if repeated samples are taken within short time period, relative to species movements, closure is easier to confirm. However, biologists may encounter situations that push the limits of the closure assumption. What if the species is not always there? What if closure is violated?

It is critical to think about closure for your species, because violation of the closure assumption can cause bias in the estimates of occupancy and detectability. And, additionally, the interpretation and inferences about the factors that influence ψ and p may be affected (Bailey and Adams 2005).

Random closure violations: does the species move in and out at random (Figure 15.2)? For example, if you use occupancy to study a large predator, such as a leopard (*Panthera pardus*), an individual may wander away from your camera trap at a specific point and come back later. **We must remember that closure is defined, in occupancy studies, at the species level**—so if other leopards remain in the area, closure is not violated (i.e., the species still occupies that location). But, if a single male represents the species on the landscape at one of your study points, closure might be violated if he leaves and no one replaces him.

So, if the target species moves away from sites during the observation periods, estimates of occupancy may be unbiased—but *only* if the species' movement (in and out of the sample unit) is random. **However, we have to redefine occupancy under these conditions**—occupancy becomes the proportion of sites used by the target species, rather than the proportion of sites occupied. "Used by the species" infers that the species used those sites during the time period of the study. "Occupied" infers continual use during the study—and the two concepts are distinct. And, the probability of detecting the species is also affected—which makes sense because the maximum likelihood estimation procedure assumes the species is present or not present throughout the study. So, the estimate of p is now a combination of the probability that the species was present at the sampling unit (see Chapter 14 for a discussion of **presence probability**) and the probability of detecting the species, given it was present.

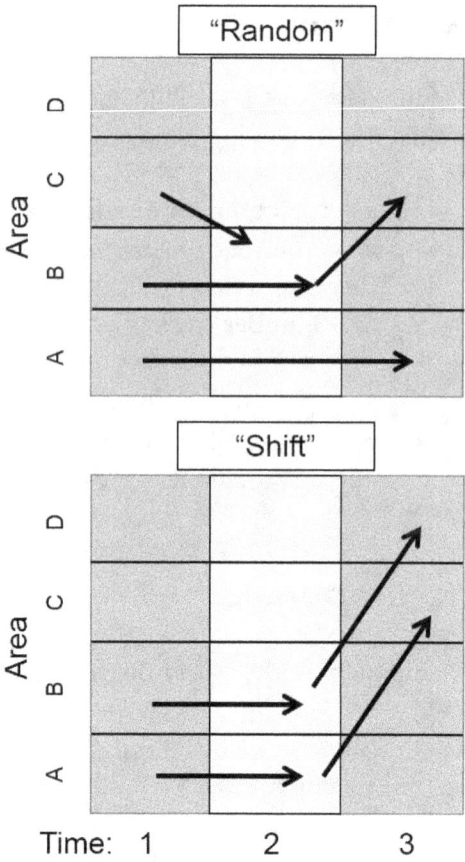

Figure 15.2: *Examples of closure violations in an occupancy study with 4 sample locations and 3 visits through time. Random violations of closure (top) occur when the species disappears and reappears at random during the study. Other violations of closure may not be random, such as migratory or other types of "shifts" (bottom) from one type of habitat to another during time frame of the study.*

If the movements that violate closure are not random—perhaps the species shifts from one section of the study area to another (Figure 15.2)—we should expect our estimates of ψ to be biased. For example, greater prairie-chickens (*Tympanuchus cupido*) shift in habitat use from contiguous grasslands during the breeding season to grasslands near crop fields during the winter in northern Nebraska, USA. If your occupancy study bridged this transition (perhaps it shouldn't?!), the species would exhibit the type of behavior shown in the bottom panel of Figure 15.2. And again, the interpretation of "occupancy" changes to the proportion of sites used by the target species during the entire sampling period. The easiest way to solve this problem is to use only the observation periods before movement occurs or to design the study with a shorter time period that does not include the period of movement.

We recommend Kendall (1999) as a good source to consult for thoughts about closure violation for all types of mark-recapture and survey study designs.

Occupancy modeling framework

MacKenzie et al. (2002) and Tyre et al. (2003) were the first to introduce ecologists to the ideas embedded in occupancy modeling. The general idea is that in a perfect world, in which $p = 1.0$, the estimate for occupancy would be the proportion of sites, s, at which animals were observed, x.

$$\hat{\psi} = \frac{x}{s}$$

But, we know that p is not usually 100%. Thus, we can state:

$$\hat{\psi} = \frac{x/\hat{p}}{s}$$

Thus, the occupancy estimate, if $p < 1.0$, will be greater than naïve occupancy.

The encounter histories for species used in occupancy modeling follow the same idea as 'capture histories' for individuals used in a mark-recapture study. For occupancy, the encounter histories indicate the success or failure of sighting the species at a site. For example, with 5 sampling occasions: 01010. The species was observed at the second and fourth occasions, but not the first, third, and fifth.

We can note that we may often have **encounter histories of 00000**. We would never have such a capture history for mark-recapture studies—as mark-recapture studies are based on a released cohort of animals, so every animal in the sample must have at least one "1" in the capture history. Of course, for occupancy studies, a 00000 indicates a site that may be unoccupied or a site at which the species (present) was missed by all surveyors.

Probability statements

Our two parameters, ψ and p, can be combined in probability statements to describe the probability of having specific encounter histories. Here are two examples, with associated probability statements:

Encounter history	Probability of having this history
01010	$\psi (1-p_1) p_2 (1-p_3) p_4 (1-p_5)$
00000	$(1-\psi) + \psi (1-p_1)(1-p_2)(1-p_3)(1-p_4)(1-p_5)$

To appreciate the probability statements above, we can use words to describe the mathematics. We use p_i to indicate the probability of being detected in period i, and that appears in each statement when the encounter history has a "1". We use $1-p_i$ to indicate the probability of not being detected, which appears in the probability statement when the encounter history has a "0". Similarly, we can use ψ and $1-\psi$ to indicate the probability that a site is occupied or not occupied, respectively.

Here, we see that for an encounter history of 00000 (note: this would apply to any encounter history without an observation of the species), we must account for <u>both possibilities</u>: (1) the species was present, but not observed, or (2) the species was not present. Mathematically, we write these two possibilities as (1) ψ $(1-p_1)(1-p_2)(1-p_3)(1-p_4)(1-p_5)$ [occupied but never detected] and (2) $1-\psi$ [not occupied].

A final note: when a site is not occupied, we do not need to include the probability of detection, p, in the probability statement as detection is not possible if the species is not available for detection. Thus, we only need to state "$1-\psi$".

Overall detection

Once we have estimated the probability of occupancy and the probability of detection, p, it is sometimes useful to estimate an **overall detection probability** (d)—the probability that the species will be detected during k surveys. We can use this formula:

$$d = 1 - (1-p)^k$$

So, for example, if $p = 0.25$, we can see that our overall probability of detection increases as we add additional replications to our survey. Ecologists may want to calculate how many surveys would be needed to ensure that there is a 90% chance of detection. Here, we see that for $p = 0.25$, we need 8 surveys to reach d=0.9. But, for p = 0.60, we only need 3 surveys (Figure 15.3).

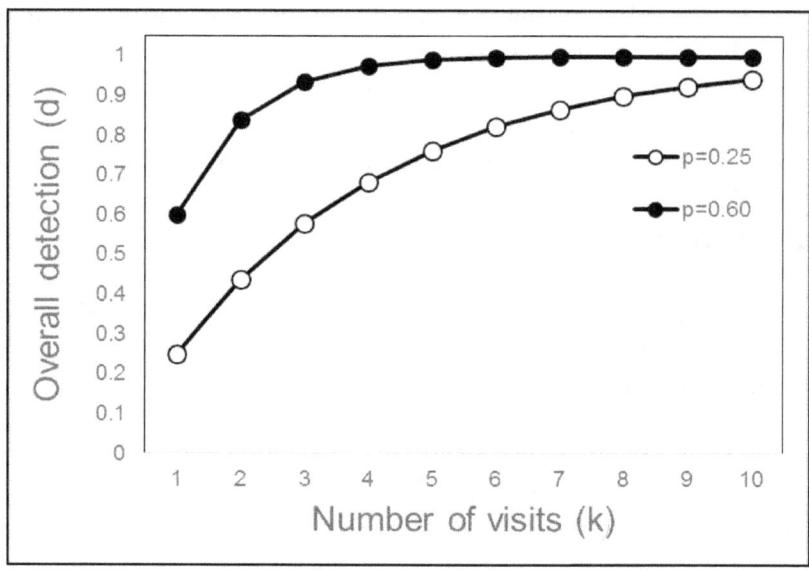

Figure 15.3: *Cumulative probabilities of detection (d) as the number of surveys conducted (k) at sample locations increases. Predictions are shown for two levels of probability of detection during a single visit (p): p=0.25 and p=0.60.*

Conclusion

Occupancy models estimate a parameter that is unique—the probability of occupancy of a site for a species. Occupancy analyses are useful when species are rare or dispersed, which results in sets of observations with many 0's. Closure, within season, is an important feature to occupancy modeling, although closure is assumed at the level of the species rather than the individual. Data collection for occupancy analysis is fairly simple and straightforward, and the data may be direct observations of animals or indirect observations through identification of scat, sign, or vocalizations.

Occupancy models can be used to address more complicated questions, which are worthy of further study. For example, occupancy can be used to assess the occurrence of species within biological communities (Dorazio and Royle 2005) or to look at co-occurrence patterns between two or more species within a community (MacKenzie et al. 2004).

References

Bailey, L., and M. Adams. 2005. Occupancy Models to Study Wildlife. US Geological Survey Fact Sheet 2005-3096. On-line: http://fresc.usgs.gov/products/fs/fs2005-3096.pdf (Accessed 3 March 2015).

Dorazio, R. M., and J. A. Royle. 2005. Estimating size and composition of biological communities by modeling the occurrence of species. Journal of the American Statistical Association 100: 389-398.

Finley, D. J., G. C. White, and J. P. Fitzgerald. 2005. Estimation of swift fox population size and occupancy rates in eastern Colorado. Journal of Wildlife Management 69: 861-873.

Kendall, W. L. 1999. Robustness of closed capture–recapture methods to violations of the closure assumption. Ecology 80: 2517–2525.

Kéry, M. 2013. Introduction to N-mixture models: Short course. EURING Technical Meeting, Athens, Georgia, USA (28 April 2013).

MacKenzie, D. I., L. L. Bailey, and J. D. Nichols. 2004. Investigating species co-occurrence patterns when species are detected imperfectly. Journal of Animal Ecology 73: 546-555.

MacKenzie, D. I., J. D. Nichols, J. D., J. E. Hines, M. G. Knutson, and A. B. Franklin. 2003. Estimating site occupancy, colonization, and local extinction when a species is detected imperfectly. Ecology 84: 2200-2207.

MacKenzie, D. I., J. D. Nichols, G. B. Lachman, S. Droege, J. A. Royle and C. A. Langtimm. 2002. Estimating site occupancy rates when detection probabilities are less than one. Ecology 83: 2248-2255.

MacKenzie, D. I., J. D. Nichols, J. A. Royle, K. H. Pollock, L. L. Bailey, and J. E. Hines. 2006. Occupancy estimation and modeling: inferring patterns and dynamics of species occurrence. Academic Press, San Diego.

Tyre, A. J., B. Tenhumberg, S. A. Field, D. Niejalke, K. Parris, and H. P. Possingham. 2003. Improving precision and reducing bias in biological surveys: estimating false-negative error rates. Ecological Applications 13:1790–1801.

For more information on topics in this chapter

Conroy, M. J., and J. P. Carroll. 2009. Quantitative Conservation of Vertebrates. Wiley-Blackwell: Sussex, UK.

MacKenzie, D. I., J. D. Nichols, J. A. Royle, K. H. Pollock, L. L. Bailey, and J. E. Hines. 2006. Occupancy estimation and modeling: inferring patterns and dynamics of species occurrence. Academic Press, San Diego.

Williams, B. K., J. D. Nichols, and M. J. Conroy. 2002. Analysis and management of animal populations. Academic Press, San Diego.

Citing this primer

Powell, L. A., and G. A. Gale. 2015. Estimation of Parameters for Animal Populations: a primer for the rest of us. Caught Napping Publications: Lincoln, NE.

Ch. 16

Double-observer methods[1]

"The way to succeed is to double your error rate."
-- Thomas J. Watson

"Double your pleasure, double your fun!"
-- Doublemint Gum

Questions to ponder:
- *How is a double-observer study conducted?*
- *What is the difference between the dependent observer method and the independent observer method?*
- *When do I get to be the primary observer?!*

Two wrongs can make a right!

The double-observer method's basic approach is to use paired observers at the same time during a survey—in an attempt to deal with the problem of incomplete detectability. At first, the use of two observers may seem odd. If we assume that one observer will miss some animals during a survey (because detection probability does not equal 100%), won't two observers be even worse? True—individually, we simply have two imperfect surveys. But, the logic of the double-observer field methods is that if we link the information from the two surveys, we can estimate the proportion of the animals missed by the individual surveys.

If we consider each survey of a location (surveyed by both surveyors) as a binomial event, we can create a capture history (see Chapter 7) to represent the success or failure of each individual. Here, the two columns represent the success or failure for sighting by the two observers:

    10—detected by the first observer
    01—detected only by the second observer
    11—detected by both observers

Surely, with this type of information, we can estimate population size? Yes, we can.

---
[1] *With thanks for content to Todd Arnold and Mary Bomberger Brown.*

Assumptions, assumptions…

The general **assumptions** of the double-observer methods:
- The population is closed during survey.
- All animals have equal probability of being detected, unless detectability is modeled by individual-specific covariates.
- No identification errors (we can clearly identify individuals into the categories needed by both of our methods).

One result of the second assumption, above, is that **availability** is not addressed by double-observer methods. That is, the method assumes all animals are available for detection.

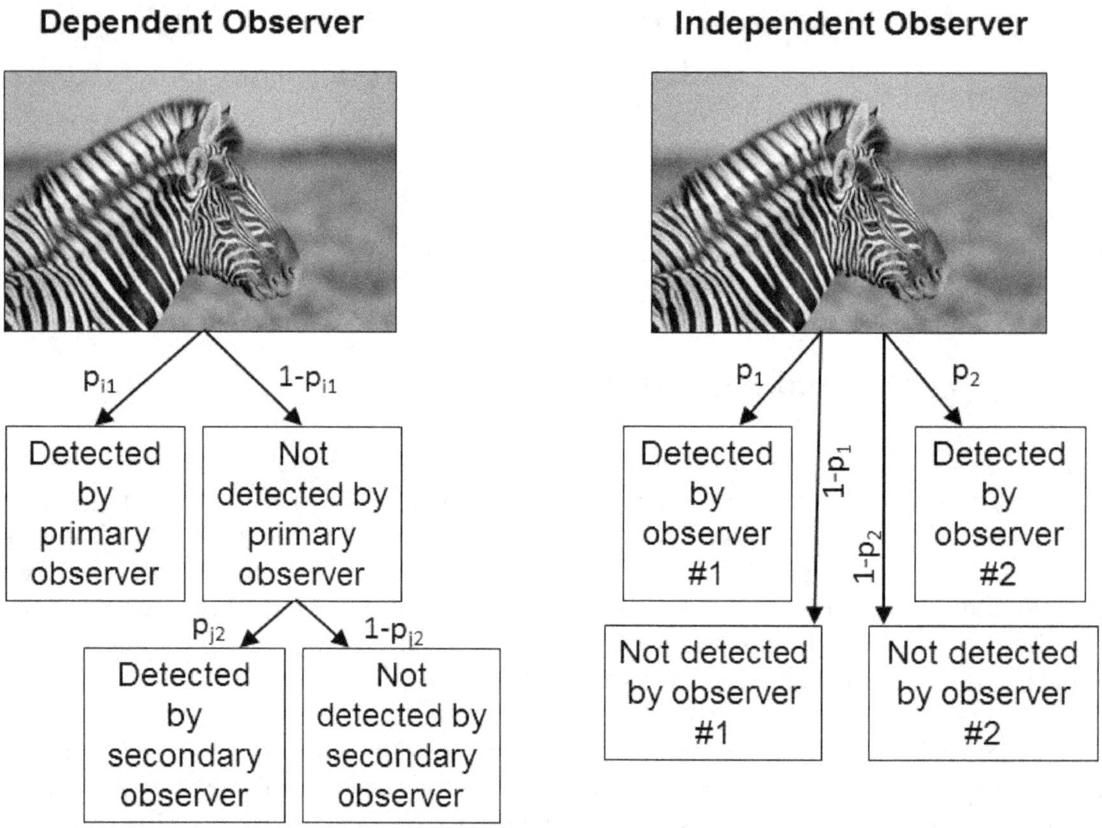

Figure 16.1: *Comparison between the timing and order of binomial events and detection probabilities for a dependent observer-type survey with double observers and an independent observer-type survey with double observers.*

Independent or dependent survey options

Double-observer methods can be used in two forms (Figure 16.1): dependent observers or independent observers. Each method has advantages and specific field methods and model structure.

The **dependent observer** method assumes that the two workers perform the survey together with interaction during the survey (Nichols et al. 2000). The method requires surveyors to switch primary and secondary roles at some point in the set of surveys, which creates a unique set of probability statements.

The **independent observer** method uses a more traditional capture-recapture approach, and the method assumes the two surveys are independent samples—one surveyor does not cue off the other surveyor (Manly et al. 1996). Because of the independence of each observation, the probability statements will be more familiar to readers accustomed (hopefully!) to encounter histories from earlier in this primer (see Chapter 8).

The differences between the two methods are illustrated in Figure 16.1 with regards to the question: *was the smallest zebra in the foreground of the photo seen?*

Dependent observer method

The field methods for the **dependent observer** method are based on a series of point counts for a specific length of time (e.g., 3-5 minutes). At each point, both observers perform specific roles as they count animals seen or heard. The **primary observer** communicates individuals that he/she sees or hears to the **secondary observer**. In addition to acting as recorder, the secondary observer records individuals that were missed by the primary observer (but seen/heard by the secondary observer; Figure 16.2). The latter point is key—the secondary observer only acts if the primary observer is observed to miss an animal (left panel, Figure 16.1). In this way, the actions of the secondary observer are dependent upon the primary observer's detections.

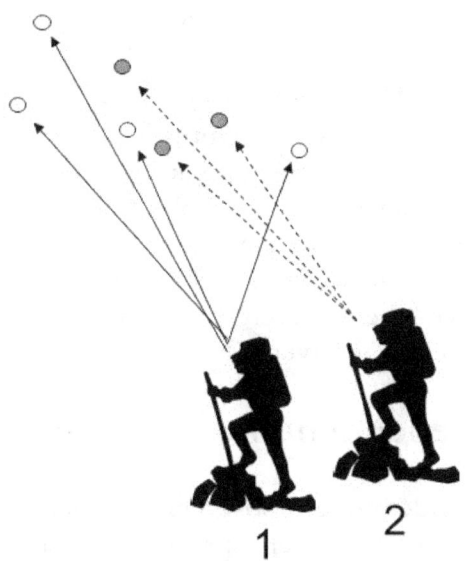

Figure 16.2: *Illustration of data collected with the dependent double observer method. Observer 1, the primary observer, communicates that they have seen a set of animals—in this case, 4 animals (shown with white dots). Observer 2, the secondary observer, then adds additional animals that he/she saw and that the primary observer missed—in this case, 3 animals (shown with dark dots). It should be noted that Observer 2 may have seen some of the animals that Observer 1 saw, but these are not recorded as Observer 2's observations.*

A critical logistical point is that the two observers **switch** roles during the survey. The switch does not have to happen at exactly the half-way point in the survey, but it is important that both observers serve in each role for a substantial number of points. In fact, surveyors may wish to switch roles at regular intervals (e.g., every-other sample point) if variables that affect detection (such as wind speed or time of day) change incrementally over the course of a survey.

We can define some statistics that will be used in the probability statements:

N_1 = true number of animals in the area sampled when observer 1 was primary observer
N_2 = true number of animals in the area sampled when observer 2 was primary observer

p_i = detection probability for observer i
x_{i1} = animals detected by observer i when observer 1 was primary observer
x_{i2} = animals detected by observer i when observer 2 was primary observer

We note that counts x_{11} and x_{22} are initial counts made by observer 1 and observer 2 in primary roles. However, counts, x_{21} and x_{12} include only additional animals that were seen by the secondary observer, but not by the primary observer—and this affects the structure of the probabilities used for the likelihood models for these counts. We can estimate the expected number of animals, E(x), counted by each observer in each role as:

$$E(x_{11}) = N_1 p_1$$
$$E(x_{21}) = N_1(1-p_1)p_2$$
$$E(x_{22}) = N_2 p_2$$
$$E(x_{12}) = N_2(1-p_2)p_1$$

Note that without switching roles, both p_1 and p_2 are inestimable. If the observers stay in their initial roles (observer 1: primary, observer 2: secondary), we will only have counts of x_{11} and x_{21}. In that case, we might think we would have information to suggest the ability of observer 1 to detect animals. For example, if observer 2 never sees/hears any animals that observer 1 missed, we might think we are gaining certainty that observer 1 has a high probability of detecting this species. Of course, it is possible that observer 2 has horrible eyesight (or hearing)! This is why we need the observers to switch roles, so that we can assess the abilities of observer 2.

Independent observer method

The **independent observer** approach is different than the dependent observer approach, because the two observers conduct independent counts of animals that are seen at a survey point. However, the two observers must record their data with methods that allow them to determine which animals were seen by both surveyors (Figure 16.3). That is, it is <u>not enough</u> to record two separate surveys—the two surveys must be **reconciled**, or compared. The tricky part of this method is the fact that observers must keep a level of independence while recording data, but must also be able to confirm that certain individuals were recorded by both observers.

One option, to maintain independence, is to use two vehicles (e.g., for an aerial or road survey). If individuals map the spatial arrangement of their observations, it should be possible, after the survey, to review data and decide which individuals were seen by both observers, and which were missed by one or the other observer. However, the lag in time may prevent a good accounting of the observations if the mapping is not done in a detailed

Figure 16.3: *Illustration of data collected with the independent double observer method. Observer 1 and Observer 2 both write down, independently, the individuals that they observe. After the survey is completed, the two observers compare notes to determine which individuals they both saw and which individuals were seen by only one observer (either Observer 1 or Observer 2).*

manner—which often can be impossible when counting a large number of animals such as migrating geese or caribou.

A second option is to use the same vehicle. Perhaps the observers keep a piece of cloth draped between them, or one sits in the front seat and one sits in the back. Observers may use some kind of repetitive motion, such as tapping their pen to the data sheet (pretending to write), to keep the other observer from keying on their writing (*"oops...there must be an animal nearby, if she's writing down data!"*). After the conclusion of the sampling at each point, or at intervals during a transect survey, the individuals may compare their data sheets and conduct a summary on the spot to determine which animals were seen by both observers. This option offers the potential for lack of complete independence (if observers key off the other), but it allows for a shorter time before summarizing the observations.

The results of an independent double-observer survey can be represented as capture, or encounter, histories, for which each line is an individual animal that is seen by either or both observers. Similar to a closed-mark-recapture experiment, we are never able to record a "00" capture history, because we do not know how many animals were missed by both observers.

We define the following encounter histories:
    10—detected only by observer 1
    01—detected only by observer 2
    11—detected by both

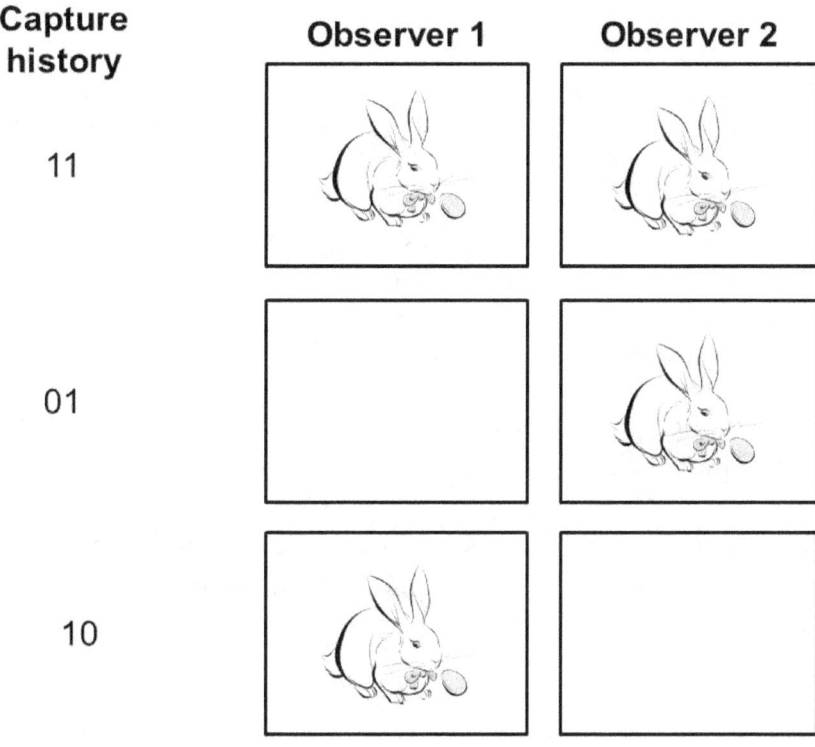

Figure 16.4: *Illustration of encounter histories for an individual animal generated for analysis of double observer data using independent observer methods. When both observers detect the individual is given a "11". If only the first or second observer detects the animal, the individual is given an encounter history of either "10" or "01", respectively.*

Using the independent method, there are only three counts (compared to four with the dependent method) to be summarized, x_{ij}, where i = observer 1 and j = observer 2 (Figure 16.4):

x_{11} = Animals detected by both observers 1 and 2
x_{10} = Animals detected by observer 1 but not by observer 2
x_{01} = Animals detected by observer 2 but not by observer 1

Even though we can summarize the data from the independent observer method as capture histories, we cannot simply use a closed-form population estimation to estimate N—the data structure is not the same! We do not have cohorts of animals that have been released and recaptured...not really.

What do we have? Both observers (because of their independence) have collected a random sample of individuals in the population. In Figure 16.5, observer #1 counted 47 animals (22+25: a random sample) and observer #2 counted 41 animals (25+16: another random sample). And, a portion of those animals (25 of them to be exact) were observed in both random samples.

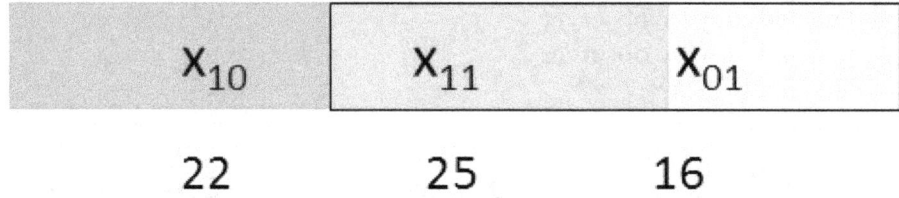

Figure 16.5: *Illustration of the random samples collected in an independent double observer survey—in total, 63 individual animals were observed (22+25+16). At left, Observer 1 was the only observer to detect 22 individuals. At right, Observer 2 was the only observer to detect 16 individuals. And, 25 individuals were detected by both observers. The method assumes that detection of any animal by either observer is a random event.*

Again, we remember the important **assumption**: all animals have equal probability of being counted by an observer. A detection is a random event.

The statistics needed for the independent observer method are:

N = true abundance (# of animals in area sampled)
p_i = detection probability for observer i

To estimate each of the p_i's, we use our random samples (Figure 16.5). If the animals seen by observer #2 ($x_{11} + x_{01}$) represent a random sample of the population, we can calculate the detection probability for observer #1 as:

$$p_1 = x_{11} / (x_{11} + x_{01})$$

Thus, p_1 is simply the proportion of a random sample (observer #2's sample) that were seen by observer #1. That may be hard to grasp, at first—but the key is to realize that the animals seen by observer #2 are not "special" animals—they are just a random sample of the population. So, how many of those animals did observer #1 detect?

Similarly, we use the random sample of animals observed by observer #1 ($x_{11} + x_{10}$) to estimate the detection probability for observer #2 (Figure 16.5):

$$p_2 = x_{11} / (x_{11} + x_{10})$$

In studies such as these, we might want to estimate the overall detection probability (p_{1+2}, the probability that an animal is seen by either observer…and thus enters the data set). We do this by first estimating the probability of both observers missing an animal:

$$P(\text{both miss animal}) = (1-p_1)(1-p_2)$$

And, therefore

$$p_{1+2} = 1 - (1-p_1)(1-p_2)$$

Now that we have our overall detection probability, p_{1+2}, we can use the standard formula (see Chapter 14) for adjusting a count, C, by detectability, β (in this case $\beta = p_{1+2}$), to estimate population size:

$$N = C/\beta$$

In the context of the independent double-observer method, we can replace C (the count) with the total number of animals detected ($x_{11} + x_{10} + x_{01}$), and we can replace β with our detection probability (p_{1+2}) to account for animals counted as well as animals that neither observer saw:

$$N = (x_{11} + x_{10} + x_{01})/p_{1+2}$$

We can analyze data gathered with the independent method for double-observer by using a modification of a closed capture method (Fletcher and Hutto 2006) such as the Huggins' or the full likelihood model (see Chapter 8) that allows the estimation of the recapture probability, c. We will have two capture occasions. And, we will estimate the probability of capture by the first observer, p_1, and the probability of capture by the second observer, p_2. We must modify the normal analysis for closed capture data by setting the probability of recapture, c, equal to the probability of capture by the second observer, so $c = p_2$.

This modification makes logical sense—we are tweaking the interpretation of the capture history to represent two observers rather than two time periods. It is intuitive to consider the first observer's 'capture' of the animal as a function of its capture probability. However, once the animal is captured by the first observer, our assumptions tell us that the animal should be no more likely, or less likely, to be captured by the second observer than an animal that the first observer missed. Hence, we set the recapture probability, c, equal to the capture probability for the second observer, p_2 as this in a sense is the only way that an animal can be "recaptured".

Irreconcilable differences?

The dependent and independent methods above require the two observers to reconcile, or summarize, their surveys—to determine who saw which animals. What if you can't do that? Or, what if you don't want to reconcile? Reconciling certainly makes the survey process more complex. But, does the failure to reconcile mean you can't use the double-observer method? The answer is: yes, but no.

Yes—traditional double-observer methods of analysis, as defined above, require reconciling. But, no—you don't necessarily have to give up the double-observer approach if you cannot reconcile. Riddle et al. (2010) proposed an "unreconciled double-observer method". It consists of two observers at the same time and place performing a survey. Rather than capture histories, the data are presented as follows for three survey points (A, B, and C):

| | Number of individuals seen | |
Survey point	Observer #1	Observer #2
A	3	4
B	5	2
C	4	4

This type of survey can be analyzed using the N-mixture methods of Royle (2004). We will cover these methods in Chapter 18, although in the "traditional" N-mixture approach, surveyors return multiple times to the same spot. Here, the multiple "visits" happen at the same time. Because the system is closed (no change in population) during the simultaneous survey, observers 1 and 2 in the example above *should* have seen the same number of animals if they both had perfect detection ($p_i = 1.0$). Of course, we suspect that $p_i < 1.0$, and for the N-mixture methods, surveyors do not have to know which of the 3 birds seen by observer 1 were also seen by observer 2 at survey point A (above)—hence, no reconciling is needed.

Potential variation in detectability

As you explore analyses with the double-observer method, you will realize that there is more variation in detection than that of the observer (perhaps based on experience or physical capabilities of sight and sound). Other factors may play a role in detection:

- **Survey-specific variation**: date, weather, methodology, etc.
- **Site-specific variation**: wetland or grassland patch size, cover type, number of additional animals to 'confuse' or distract, etc.
- **Observation-specific variation**: size of study species, sex, group size

Any of these factors can be modeled as covariates using most software packages available for double-observer estimation.

Conclusion

Double observer surveys are a relatively simple way to account for incomplete detectability. Double observer studies can be completed in a single day, because the method used to estimate detection involves the addition of a second observer to a single survey—rather than adding additional surveys (e.g., *N*-mixture methods, Chapter 18). However, double observer methods do involve some logistics and reconciling, so you might decide that it is easier to use one observer and record distances to sighted animals to estimate probability of detection (Chapter 19)—especially for surveys of very large groups of animals that are hard to reconcile between two observers' notes.

The double observer methods link data recorded by two observers, and we are able to estimate the detection probability for each observer. Two methods, the dependent observer method and the independent observer method, are possible to use. Both methods require accurate notes to be taken during the survey to determine whether the observers were detecting the same individual animals or not. Forcey et al. (2006) evaluated the dependent and independent methods for surveys for songbirds; they recommended the use of the dependent survey methods to reduce the amount of time needed to reconcile independently-gathered data. Your study species and survey design may be unique, and we recommend that you consider the information provided by table at the end of Chapter 14 as you select a survey method.

References

Fletcher Jr, R. J., and R. L. Hutto, R. L. 2006. Estimating detection probabilities of river birds using double surveys. The Auk 123: 695-707.

Forcey, G. M., J. T. Anderson, F. K. Ammer, and R. C. Whitmore. 2006. Comparison of two double-observer point-count approaches for estimating breeding bird abundance. Journal of Wildlife Management 70: 1674-1681.

Manly, B. F., L. L. McDonald, and G. W. Garner. 1996. Maximum likelihood estimation for the double-count method with independent observers. Journal of Agricultural, Biological, and Environmental Statistics 1: 170-189.

Nichols, J. D., J. E. Hines, J. R. Sauer, F. W. Fallon, J. E. Fallon, and P. J. Heglund. 2000. A double-observer approach for estimating detection probability and abundance from point counts. The Auk 117: 393-408.

For more information on topics in this chapter

Conroy, M. J., and J. P. Carroll. 2009. Quantitative Conservation of Vertebrates. Wiley-Blackwell: Sussex, UK.

Citing this primer

Powell, L. A., and G. A. Gale. 2015. Estimation of Parameters for Animal Populations: a primer for the rest of us. Caught Napping Publications: Lincoln, NE.

Anthony Pagano (back) and Andrew Dath (front) conduct a double observer-type waterfowl survey near Cando, North Dakota, USA. Photo by Matt Pieron, used with permission.

Ch. 17

Removal sampling[1]

"[The Man] will go on taking and taking, until one day the World will say, 'I am no more and I have nothing left to give.' "
-- Old Story Teller (Apocalypto 2006)

Questions to ponder:
- *How is a removal study conducted?*
- *How is the removal method similar to a closed mark-recapture study?*
- *What is the difference between removal methods with regression and removal methods with mark-recapture methods?*

Sample until you drop?

The methods for removal sampling are based on a simple idea—if you capture samples of animals in quick succession and remove the captured individuals each time, you will eventually sample every animal. And, each successive sample will have fewer animals than the previous, as there are fewer animals that remain to be sampled.

As an example, consider a small pond full of ornamental fish (such as koi, or common carp [*Cyprinus carpio*]) in a garden. If you use a net to capture fish, you will capture several on the first attempt. You remove the fish and place them in a holding tank before sampling the pond again. Now, there are fewer fish in the pond than when you started, so you should catch a few less fish on your second attempt. The process continues with each swipe of your net. Eventually, if you keep sampling the pond, you will capture all of the fish. And, if you counted them as they went into the holding tank, you would know the population size of the pond.

However, let's assume you are somewhat lazy, or more likely, let's assume that complete removal isn't practical! Do you really want to spend so much time sampling the pond—until you are certain you have caught every fish? Or could the same information be ascertained if you stop after 4-5 samples, for example? Would the *pattern* of the sample sizes of your captures allow you to estimate how many fish are in the pond?

---

[1] *With thanks for content to Gary White and Mark Pegg*

The Zippin Depletion Method

P. A. P. Moran (1951) was an early proponent of a "mathematical theory for the estimation of the total population of an animal in a given region from the records of a series of trapping by simple traps (each of which killed the animal)…" In fact, capture-removal methods were a standard in early mammal and fisheries research. Moran's influence has largely been forgotten, because small mammal and fisheries biologists eventually named the method the "Zippin Depletion Method", after Calvin Zippin who published two useful papers for the estimation of small mammal population size (Zippin 1956, 1958).

Removal sampling is still used, perhaps most commonly, by fisheries biologists. For example, electrofishing is a capture technique that lends itself well to the removal method—and the classic use of removal sampling has been to sample a stream that has been blocked at points upstream and downstream from the sample area to prevent movement out of the area. A backpack-mounted device is used for electrofishing, and the first sample of temporarily stunned fish are removed from the stream and placed into a holding tank. However, some fish remain in the stream. More electricity and another pass through the section of the stream, and a second sample is removed from the population. And, the sampling continues. There is no reason to mark the fish as fish have been removed from the population, which makes the method quick to accomplish in the field. Of course, you could mark the fish and return them to the stream—then you would only count the unmarked fish in the second sample—ignoring the previously marked fish.

Removal methods have broad application to other taxa and other logistical methods for sampling with 'removal'. Small mammals can be captured with snap traps (a form of 'recovery' from which the animal cannot be released) or live traps. Removal sampling may also be effective in visual or auditory surveys of birds, frogs, or other animals. Typical point count methods are modified to include time intervals within the time spent at a point. For example, you might break a 4-minute point count into 4, 1-minute segments. The observer keeps track of how many new animals are seen or heard in each time interval. Animals seen in the previous time interval are ignored---just like they would be during a regular point count (you only count animals once!).

For example, during an auditory survey of frogs, you might record the following:

>   Interval 1: 10 frogs
>   Interval 2: 5 frogs (not detected previously)
>   Interval 3: 4 frogs (not detected previously)
>   Interval 4: 1 frog (not detected previously)

It should be clear from the trends you can see in the data that you are getting close to observing all of the frogs in this patch of habitat—there are, perhaps, about 20 frogs (10+5+4+1).

Regression methods for removal

Even though Zippin (1956, 1958) described the removal method in terms of binomial events, he also provided a "regression method of estimating population size from removal data." Perhaps due to ease of calculation, early biologists generally turned to the regression method, which was an innovative statistical method to estimate the population size in the area sampled. As subsequent samples were drawn, the catch in the current time period was recorded and also the cumulative catch at the end of the current time period—as in this example:

Time period	Catch	Cumulative catch
1	82	82
2	33	115
3	22	137

Even after 3 time periods, a couple things are clear: (1) we haven't caught all of the animals in the area yet, and (2) we are beginning to get close to the total number of animals, because our sample sizes of captures are dropping significantly.

We can create a plot of our data—with the cumulative captures on the y-axis and the captures in time period i on the x-axis (Figure 17.1). If we perform a standard linear regression analysis, we will be given the y-intercept. And, the y-intercept is the statistic in which we have interest—it represents the estimate of the population size at our sample point, 150 individuals are estimated for this example. The logic of this approach is that the y-intercept is the predicted cumulative catch when we capture 0 animals during time period i—or, the theoretical point at which no more animals are left to catch (Figure 17.1).

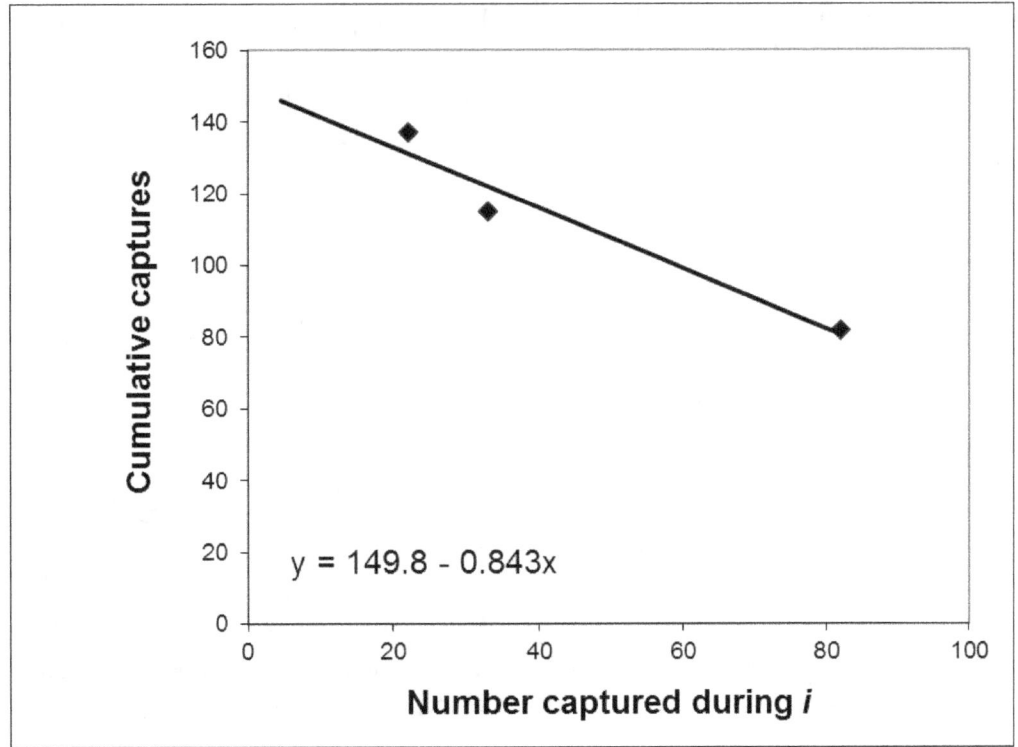

Figure 17.1: *The Zippin deletion regression approach, with cumulative number of animals captured on the y-axis and the number captured during each interval, i, on the x-axis. The y-intercept serves as the estimate for the population size, N. Here, the population size estimate is 149.8 (SE=10.6).*

We can obtain a 95% confidence interval for our estimate of N by using the SE of the intercept (95% CI = +/- 1.96*SE) from our regression analysis. It seems like a slick method (and, conceptually, it is very impressive!).

However, there is a potential problem. Performing a regression on 3, 4, or even 5 points is highly questionable (e.g., to obtain growth rates for polar bear cubs). And, if we did a regression using 3-5 sample points, we would expect large standard errors and large confidence intervals. And, this is also true of the standard regression methods for removal sampling.

A second problem is that the regression approach, by its linear nature, with a constant slope assumes a constant probability of capture, p. By now, you should anticipate that capture probabilities may vary according to many factors, and the use of regression does not allow you to test for variation in p across your sampling periods.

We should also point out a feature of the regression method for removal—the capture probability (p, the portion of the population that will be captured in a given sampling effort, i) affects the slope of the regression line. In the simulation result shown in Figure 17.2, two sampling methods are used to sample a population of 1000 animals with removal sampling. The netting method used on the left results in a 50% capture probability (p)—so, more animals are captured and the depletion is achieved more rapidly. The slope, therefore, is shallow. However, if capture probability, p, is only 5% (the example on the right), fewer animals are captured, and the slope is steep. Both give the estimate of N~1000, but the SE of the estimate for the intercept is larger for the figure at right (SE = 34.2) than the figure at left (SE = 0.3), because the lower capture probability leads to a smaller range covered on the x-axis. Thus, we have more confidence in our population estimate for the study represented by the figure on the left, with higher capture probability (p=0.50).

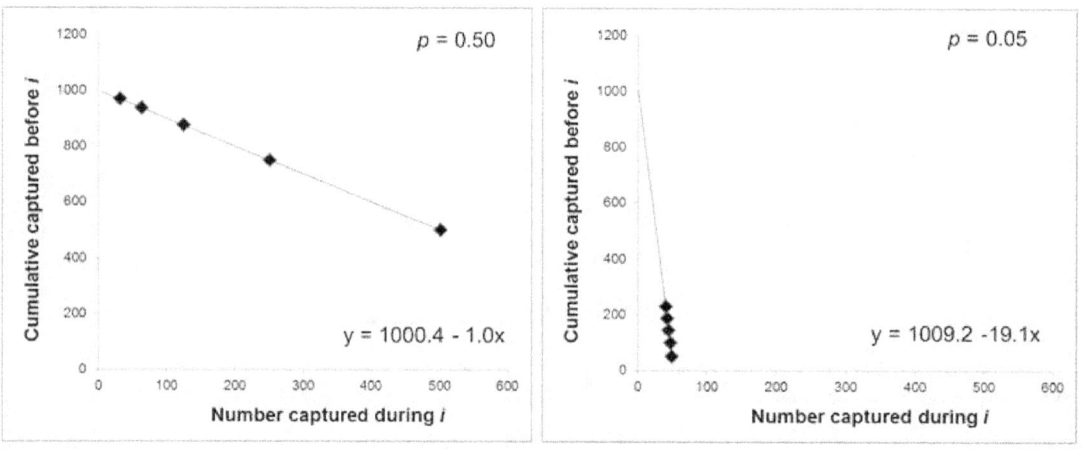

Figure 17.2: *Comparison of regression analyses for a removal experiment for a simulated population of 1000 animals sampled with capture probability of p=0.50 (left, $\hat{N}$ =1000.4, SE=0.3) or p=0.05 (right, $\hat{N}$ =1009.2, SE=34.2).*

Modern removal methods

It is advantageous to think about removal methods from a mark-recapture framework (Otis et al. 1978). In fact, you will recognize that removal sampling is exactly like closed mark-recapture (see Chapter 8) sampling to estimate population size—except for the important fact that we do not release any animals for recapture. Thus, **the probability of recapture, c, is 0**.

We can use modern maximum likelihood-based methods to estimate the population (N) from removal methods by thinking of each sample as a binomial event. Let's consider a simple case in which the survey is divided into two samples:

N: the population size within 'reach' of the sampling gear
x_i: the number of new animals (never observed previously) counted in time i, t_i

Thus, we can estimate the expected number of animals captured in each time period, $E(x_i)$ as:
$E(x_1) = Np_1$
$E(x_2) = N(1-p_1)p_2$
$E(x_3) = N(1-p_1)(1-p_2)p_3$

Note that the mark-recapture framework allows unequal "catchability" or encounter rates during the experiment, which may be advantageous. Removal data can be analyzed in any mark-recapture software program that allows the user to modify the probability of recapture, c. The user must set $c = 0$ for removal methods.

Then, capture histories can be created for individuals sampled. To match the 3-sample removal data used in our previous regression example, the capture histories would be:

100 -- 82 individuals with this history
010 -- 33 individuals with this history
001 -- 22 individuals with this history

Removal sampling assumptions

Removal sampling has two very simple **assumptions**, in addition to the assumptions inherent in all closed-population, mark-recapture studies (see Chapter 8). First, we assume a closed population within a sampling area during the sample duration (across intervals). And, we assume that there is no double-counting of individuals (during and across intervals). This assumption is more easily violated when animals are marked and returned to their habitat or visual/auditory counts are occurring.

Practical considerations

Removal sampling is conducted during one primary sampling period (e.g., 1 visit to a stream), but the method requires several sub-samples, and that effort may create a longer sampling event than a single count-type survey. If used for visual surveys (e.g., grassland or forest birds), removal sampling requires additional logistics of timing the intervals of the surveys (perhaps with a stop watch). The number of animals and species at a given spot may make removal sampling impractical—for example, in many locations in Colombia's forests, there are over 200 species of frogs with large populations. Keeping track of who called and where would drive even the sturdiest person insane and would likely produce erroneous population estimates! Similarly, in any high density area, such as trying to count red-winged blackbirds (*Agelaius phoeniceus*) in a wetland in the central US, the logistics of keeping track of animals that are moving within the sample area during a survey would prove to be impossible. However, using the method to count calling grassland birds may prove very possible.

Adjustments can be made to sampling protocols to use removal sampling. For example, biologists could limit the number of species that are being counted to a list of target species. The time of the survey could be limited—instead of a 10 minute survey with 5, 2-minute intervals, perhaps using a 3-minute survey with 3, 1-minute intervals could be used for red-winged blackbirds in the example above. Errors in double-counting increase with interval length and replication of intervals. Special data sheets can be used to keep track of new individuals seen in each time interval (Figure 17.3).

Removal methods can be applied or adjusted to many types of animals. However, to be useful, there must be a large initial capture event, and there needs to be a pattern of declining numbers of captured animals. The method, for example, would not provide good results if a small sample was taken each time from a large population of 10,000 animals. No noticeable decline in the size of the sample of captured animals would be observed in this situation—the population is not being depleted!

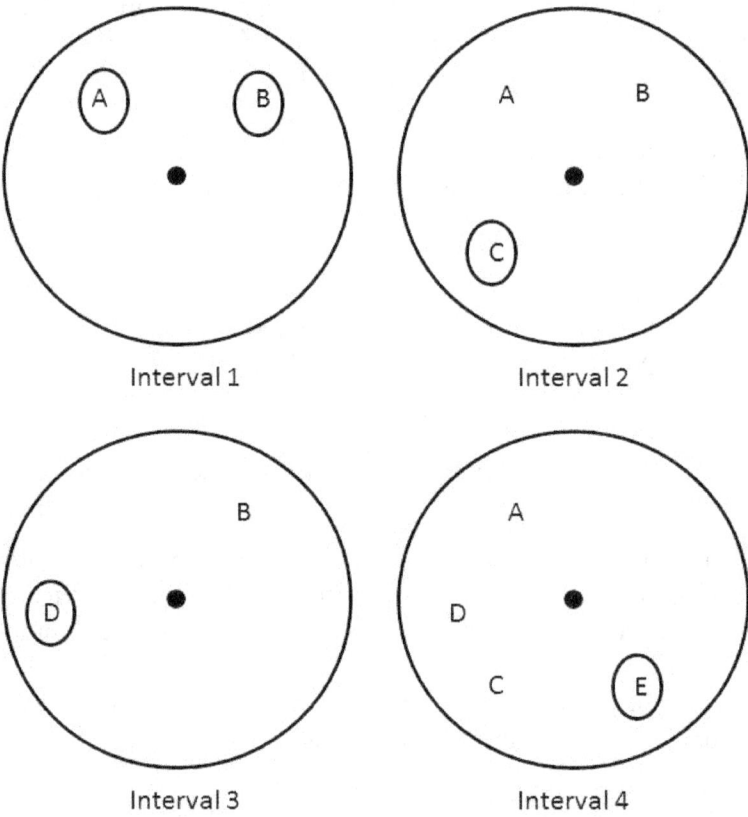

Figure 17.3: *A data sheet for use at a point count for birds that allows summary of data for removal analyses. The data collected during each time interval is plotted on the map of the point count (observer at center dot). Unique letters indicate unique individuals seen during interval. New observations are circled.*

Conclusion

Removal methods are some of the oldest methods in fisheries and wildlife ecological research. Only certain situations allow for use of the removal method, because significant progress towards depletion of the local population must occur during sampling. In addition, logistics for holding tanks (e.g., for fish) must be available; in the case of small mammal removal with snap traps (causing death), animals are not released back to the population, which is not feasible for species of conservation concern. However, the method may be adapted for use in visual/auditory surveys and mark-recapture studies that include release of live animals. Regression methods provide quick-and-easy estimates of population size, but they assume equal capture probability for individuals in the population during the study. Modern mark-recapture methods are preferred, and allow investigation of variability of capture probabilities.

References

Apocalypto. 2006. Director: Mel Gibson. Icon Entertainment International, Film.

Moran, P. A. P. 1951. A mathematical theory of animal trapping. Biometrika 38:307–311.

Otis, D. L., K. P. Burnham, G. C. White, and D. R. Anderson. 1978. Statistical inference from capture data on closed animal populations. Wildlife Monographs 62:1–135.

Zippin, C. 1958. The removal method of population estimation. Journal of Wildlife Management 22:82–90.

For more information on topics in this chapter

Conroy, M. J., and J. P. Carroll. 2009. Quantitative Conservation of Vertebrates. Wiley-Blackwell: Sussex, UK.

Williams, B. K., J. D. Nichols, and M. J. Conroy. 2002. Analysis and management of animal populations. Academic Press, San Diego.

Citing this primer

Powell, L. A., and G. A. Gale. 2015. Estimation of Parameters for Animal Populations: a primer for the rest of us. Caught Napping Publications: Lincoln, NE.

Fisheries biologists use nets and electroshocking to estimate population size in a local area of a lake with removal sampling. Photo of Mark Pegg and University of Nebraska-Lincoln students by L. Powell.

N-mixture models[1]

"There is no great genius without a mixture of madness."
-- Aristotle

Questions to ponder:

- *How do N-mixture models differ from occupancy models?*
- *How do I gather data for analysis with N-mixture models?*
- *Why are these models called "mixture" models?*
- *How important is the closure assumption for N-mixture models?*

N-mixture model basics

Royle (2004) introduced the concept of *N*-mixture models, which are used to estimate abundance from spatially and temporally replicated surveys. In fact, many small-scale studies or large-scale monitoring efforts are replicated spatially because the ecologist often has a goal to characterize spatial variation in abundance. And, these studies often have multiple visits over time. The end-result of these studies is a set of data, $y_{i,k}$, a set of observed counts from each site i and each visit k.

The essence of the study design is shown in Figure 18.1 for three sites that are sampled over 4 time periods. In Chapter 15, we saw a similar design for occupancy methods, but the response values for the data were either 1's or 0's (species observed or not observed). Here, the response data is the number (typically counts) of individuals seen at a point.

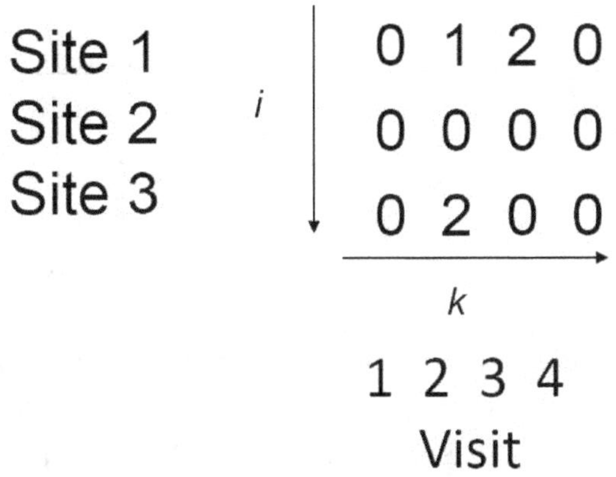

As with previous survey methods, *N*-mixture models are designed to account for incomplete detectability. And, we can see evidence of detectability problems in our data. For Site 1, we can see that the minimum number of animals is 2. We did not see any individuals during visit 1 and 4, and only 1 during visit 2. So, detectability is not perfect. The patterns in our data might lead us to

Figure 18.1: *Example design for a study using N-mixture methods with 3 sample sites (I) and 4 visits (k) to each site over time.*

---

[1] *With thanks for content to Max Post van der Burg, Therese Donovan, James Hines, and Marc Kéry*

suspect that there may be more animals than two at Site 1. And, that is the basic idea of mixture models—using the patterns of how many animals were seen at a site to estimate the actual number of animals at that site.

Previous to the advent of N-mixture models and more rigorous methods to estimate abundance at sites, biologists used the **mean count** (for Site 1, mean count during 4 surveys = 0.75) or **maximum count** (for Site 1, maximum count = 2) to represent the abundance at each point. We sometimes refer to that representation of abundance to compare to the response at other locations as **relative abundance**. But, these types of counts will obviously under-estimate the true abundance if $p \neq 1.0$. Of more concern is the fact that raw counts taken in dense habitats may be done under conditions of lower detection probability than raw counts in sparsely vegetated areas. So, the relative abundance metrics may be biased at best and not comparable at worst.

A mixed landscape

In Chapter 15, we defined the **latent state** of our sampling point as the true state of existence that is hidden or concealed. Specifically to survey sampling, the latent state is the true number of animals at a given sampling site. We conduct surveys to try to estimate how many animals there are, as we do not know the latent state, by definition.

Occupancy models (Chapter 15) assume a binomial distribution of both the latent state and the observations (1's and 0's for present or absent). N-mixture models, in similar fashion, assume a binomial distribution for the observations (each animal is either seen or not seen). But, the latent state is assumed to be distributed as Poisson. The Poisson distribution is often used for "count data", and the Poisson distribution can be used to describe the probability of x individuals at a site, given an average number of λ individuals at each site.

Before we look at the distribution in more detail, we can take an example of raw data (summarized as encounter histories) collected at 5 hypothetical sample points during 3 sampling time periods:

Point	Latent State	Data (Encounter Histories)
1	4	0-0-4
2	2	0-1-1
3	3	0-0-0
4	0	0-0-0
5	9	5-6-2

Point 1 had 4 individuals near it (the unknown latent state), and all 4 were seen—but only on the third survey. The two individuals (latent state) at Point 2 were never seen as a group—only one individual seen during the last two time periods. The three individuals at Point 3 were never seen. Point 4 had no individuals (latent state), so none could be detected. And, the observer only saw a portion of the 9 individuals at Point 5 during each of the three surveys.

***N*-mixture analyses assume closure of individuals during all survey periods** (all individuals are present at the same spot through time, which differs from the species-level closure assumption for occupancy analyses—see Chapter 15), so we can begin to see that our maximum likelihood estimate for detection probability will not be 100% for our 5-point example above. And, we can further guess that the estimate for N at each point will be greater than the number of individuals seen during the survey periods.

A bit about the Poisson

It is important to understand the general idea of the Poisson distribution, just as we have described the binomial distribution in previous chapters. The Poisson distribution has one parameter, λ. And, we can define λ as the average number of animals at each location (unknown—the latent state).

The probability density function f(x) for the Poisson describes the probability of having x animals at a site, when the average number of animals at a site is λ (Figure 18.2). You might conceptualize this by asking the question—*if I go to a random point in my study area, how many animals am I most likely to see?* The peak of the probability density function will give you your answer, and you can see how probable other counts are predicted to be.

$$f(x) = \frac{e^{-\lambda} \lambda^x}{x!}$$

The Poisson predicts, by its shape and function, that there will be (with much higher chance) λ individuals at each point we sample. That is, the most common number of individuals at sites should be equal to the mean, λ. In contrast, there is a very low probability of seeing a number of individuals that is very far from the mean (either lower or higher). Of course, the number of individuals at a site is not known, and we are trying to estimate λ.

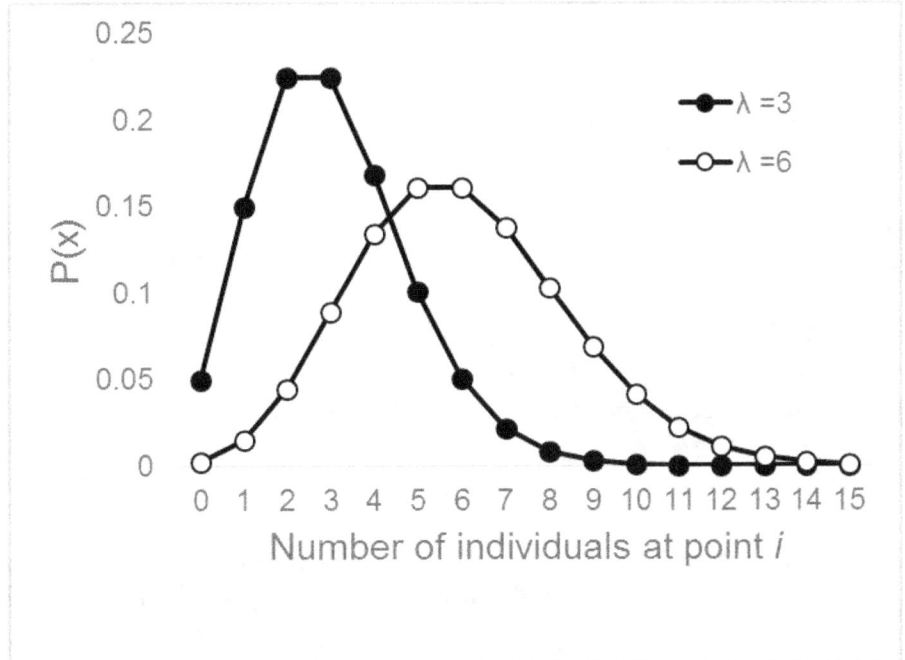

Figure 18.2: *Probability density functions, P(x), describing the distribution of animals at all sample points when the average number of individuals (λ) at all points is either 3 or 6.*

As an example, Figure 18.2 shows two probability density functions—one for λ=3 and one for λ=6. The peaks of the functions maximize at either 3 or 6, so the functions show that it is most probable for sites to have local numbers of animals of either 3 or 6. Although it is possible to have 10 animals at a site when λ=3 or λ=6, but it is very unlikely.

Mixture sampling might initially seem similar to closed mark-recapture methods, because we estimate N. However, mixture modeling is conducted in a completely different context than mark-recapture methods—the underlying distribution of animals on the landscape is assumed to vary with some pattern during mixture-based analyses (Figure 18.3, right). And, mixture-based analyses use point-specific samples to estimate the variability in N. In contrast, mark-recapture methods use a cohort of marked animals within a study area to estimate the population size for a study site, and N is assumed homogeneous for the entire site (Figure 18.3, left).

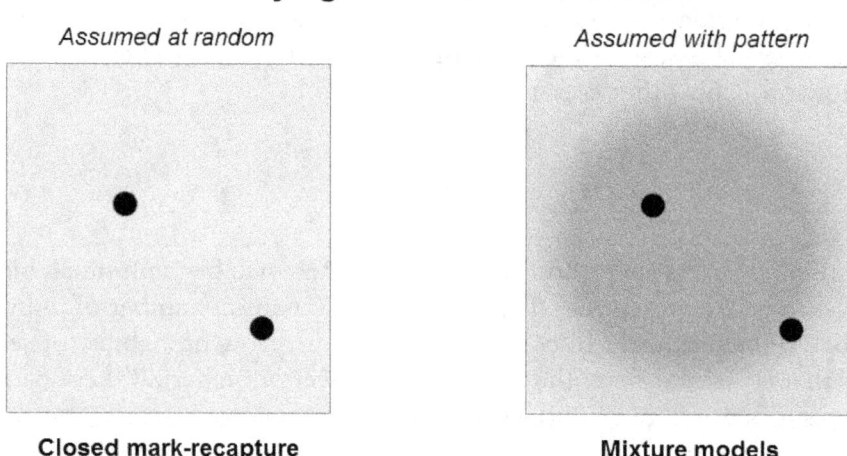

Figure 18.3: *Comparison of the distribution of animals under closed mark-recapture (left) and mixture models (right). The black dots represent sample points. The transition in shading on the right figure indicates changes in population density within the area, while a constant distribution of animals is assumed for closed mark-recapture analyses.*

To further explore this difference, we can transition from a graphical description to a statistical definition—and we will discover the reason for the name, **mixture model**.

Statistically, we can define $E(x_i)$ as the expected count (of your data). In closed mark-recapture models, the expected count of captured animals from a set of traps or nets is a function of the population size, N, and the encounter rate, p. We can write this as:

$$E(x_i) \sim f(N, p_i)$$

In contrast, N-mixture models are structured such that N is a function of two distributions. First, we can state, similar to mark-recapture models, that our expected count at a sample site is a function of the population size at the site, N, and the encounter rate, p. So, we still have:

$$E(x_i) \sim f(N_i, p_i)$$

However, we expect local abundance (at each sample point) to vary across our study's surface, and we define local abundance as λ_i. Thus, the population size, N, is a function of the cumulative local abundance estimates, so

$$N \sim g(\lambda_i)$$

To visualize the mathematics of how we assess a population with N-mixture models, consider Figure 18.4. We see that the local abundance (λ_i) varies spatially (represented by size of light-colored circles)—perhaps a covariate in our analysis would describe why λ_i is larger at some points and smaller at other points. And, we see that the observations (x_i, dark-colored circles) at many locations are incomplete ($p<1.0$). N-mixture models have become popular methods for analysis of survey data, because the analysis allows a focus on 'ecological processes' or habitat associations rather than on the estimation of density.

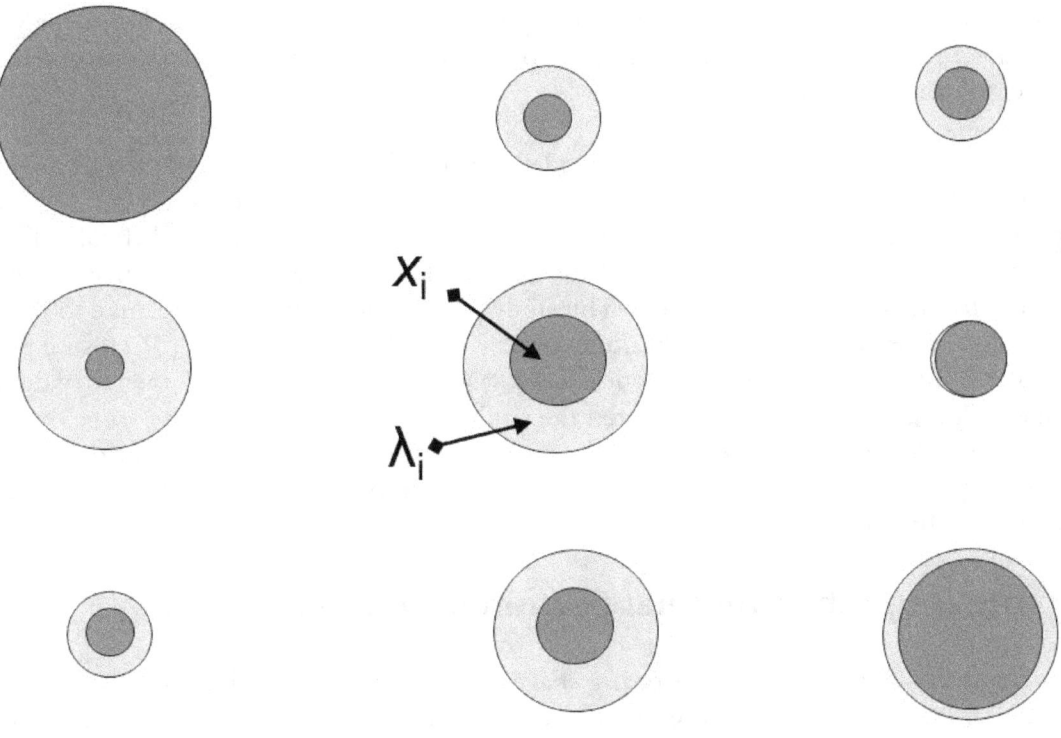

Figure 18.4: *Illustration of the concept of local abundance (λ_i) used in analyses with N-mixture models. Survey sample locations are shown as a grid of points. The general structure for these analyses can be seen as 'metapopulations' each with a local abundance (lighter polygons, λ_i) of animals, of which only a portion are observed during surveys (darker polygons, x_i).*

N-mixture model assumptions

- The distributions assumed (binomial and Poisson) are true
- Observers do not double-count at a sample point within a sampling session
- The number of animals inhabiting one site is random and independent of the number of animals at other sites
- Population is closed between surveys
- Detection is constant across sites, unless modeled by covariates
- All N_i individuals at occasion k have same detection probability p_{ik}

Research designs that use *N*-mixture models for analysis do not require identification of individuals between occasions, as no matching of specific individuals is required (see Chapter 16 on double-observer methods). However, the 'identity' of individuals within a sampling session must be known to the extent that surveyors must keep track of individuals and not double-count during the survey. Because of the assumptions about detection probability, we can see that variation in detection because of distance from the point is not incorporated explicitly into *N*-mixture models (see distance sampling in Chapter 19). Therefore, some ecologists may choose to use fixed-radius sampling to minimize the problem with individuals seen, or not seen, at distances farther from the point.

The closure requirement for *N*-mixture modeling is much more severe than for occupancy modeling. If the closure assumption is violated, estimates of N may instead refer to a super-population in some manner, although interpretation of such an estimate is difficult. To ensure closure is not violated, ecologists should use careful planning during the design stage with special attention to the duration of the study. The study period should be short, relative to the movement dynamics expected of the study organism in the target system. If closure problems are realized during or after data collection, ecologists have three options—they may discard data from early or late time periods to shorten the study, they may attempt to use some of the "open" mixture models that are now available (but see our caution below), or they may decide to switch to occupancy modeling to take advantage of the more relaxed closure assumption at the species level, rather than the individual.

Mixing it up: the binomial portion of the likelihood

***N*-mixture models** *are literally a **mixture** of different distributions.* The likelihood statement for the binomial portion of the *N*-mixture model, below, describes the probability of seeing y animals, given a detection probability, p, (probability of detecting an animal given it is present and available), and a given population of animals at the site (N). The data we collect is y in this likelihood description. Our software will use maximum likelihood estimation to estimate p given our set of samples.

$$f(y \mid N, p) = \binom{N}{y} p^y (1-p)^{N-y}$$

For a mental exercise, let's assume that $N = 10$. If our probability of detecting an individual is high (e.g., p=0.9), do you think it is more likely that we would see 2 animals or 8 animals of the 10 that are possible to see?

Logically, it makes sense that if we have high detection probabilities, it should be <u>more likely</u> (our working definition of "maximum likelihood") to see 8 of the 10 animals than it would be to see 2 of the 10 animals. And, it is easy enough to check it (the more we use MLE methods, the less scary they seem!):

$$f(y\,|\,10,0.9) = \binom{10}{2} 0.9^2 (1-0.9)^{10-2} = 0.0000003645$$

$$f(y\,|\,10,0.9) = \binom{10}{8} 0.9^8 (1-0.9)^{10-8} = 0.193710245$$

Mixing it up: adding the Poisson portion

The entire likelihood statement for the N-mixture model is shown here:

$$p(y_{ik}\,|\,\lambda, p, N_i) = \prod_{i=1}^{R} \left(\prod_{j=1}^{S} Bin(y_{ik}\,|\,N_i, p) Pois(N_i\,|\,\lambda) \right)$$

Where

y_{ik} = observed counts from site i and visit k
N_i = site-specific unobserved ('true') abundances
p = the <u>overall</u> detection rate (estimated from data)
λ = overall unobserved mean density per site

To understand this likelihood statement, we start on the left side of the equality (=) with "$p(...)$". We read this as the **probability of a count at a site and time, given a mean abundance (λ), detection probability (p), and population at the site (N)**.

On the right side of the equation, we can see a binomially distributed portion ("*Bin*"). This portion of the likelihood describes the binomial-distributed **probability of detecting y animals, given N animals at the site and a detection rate (p)**.

The Poisson portion ("*Pois*") of the statement describes the Poisson-distributed **probability that there are actually N individuals at site i, given that the mean abundance across all sites is λ**.

When we throw all of this together, we can see that we should weigh the probability of observing a certain number of animals over k visits (binomial portion) by the probability that the site contains N animals (Poisson portion). Our software will use maximum likelihood-based methods to estimate values of λ, p, and N which are the most likely, given your observations (y).

Adaptive radiation of mixture modeling: a caution

The use of mixture models is rapidly advancing in ecology. And, recent years have seen a large increase in the number of modifications to the basic methods we describe here. Examples include:

- Mixture methods that allow spatial autocorrelation between sample points (Royle et al. 2007a, Post van der Burg 2011)—the methods we have described in this chapter, above, assume independence among observations from different sample points.

- Mixture methods to estimate species richness (Dorazio and Royle 2005, Royle et al. 2007b, Royle and Dorazio 2008).

- 'Open' N-mixture models that allow violations of the basic closure assumption (Dodd and Dorazio 2004, Chandler et al. 2011, Dail and Madsen 2011).

The reader will no doubt be able to find additional modifications with a quick literature search. We caution you with regard to jumping to the newest, most complicated model structure—as with other methods in this book (e.g., multi-state models, robust design models), the complicated methods literally become complicated by adding parameters. "Open" N-mixture models have added recruitment and extinction-type parameters, for example—and more parameters to estimate requires you to collect more data, all else equal! We encourage the reader to design your study to use the simplest method possible to answer the question(s) you have. We caution against turning to complicated "open" models, for example, as a remedy for closure violations that could have been prevented with simple changes to the study design.

Conclusion

N-mixture models can be useful for count-surveys that result in a large number of 0's for a species, which can be a result of small populations or low detection rates. The lack of data, in these situations, may make it difficult to use other "standard" approaches such as distance sampling, double-sampling, removal methods, or mark-recapture. N-mixture models perform better in these situations because there are multiple visits, during which an animal may be detected that was not detected previously (resulting in a 0-1-2-0 encounter history, perhaps). If species are extremely rare (with many unoccupied sites that result in 0-0-0-0 encounter histories), N-mixture models may not be able to estimate parameters—the biologist might consider the use of occupancy models as site occupancy is often more desired for rare species than local abundance predictions. However, stay true to your initial goals for your research!

N-mixture models do have a closure assumption that should not be violated. And, the multiple visits, through time, to multiple sites do require planning and consideration of the effort needed. Other methods, such as distance sampling, removal, or double-observer can provide estimates of either density or abundance with one temporal survey, but the survey must result in large number of detected individuals. And, distance, removal, and double-observer are intensive type of surveys compared to simply counting individuals for use with N-mixture models. We encourage you to evaluate the information in the table at the end of Chapter 14 to assist with your research design and planning.

References

Chandler, R. B., J. A. Royle, and D. I. King. 2011. Inference about density and temporary emigration in unmarked populations. Ecology 92:1429-1435.

Dail, D., and L. Madsen. 2011. Models for estimating abundance from repeated counts of an open metapopulation. Biometrics 67:577-587.

Dodd, C.K. Jr., and R. M. Dorazio. 2004. Using counts to simultaneously estimate abundance and detection probabilities in a salamander community. Herpetologica 60:468–478.

Dorazio, R. M., and J. A. Royle, J. A. 2005. Estimating size and composition of biological communities by modeling the occurrence of species. Journal of the American Statistical Association 100: 389-398.

Kéry, M. 2013. Introduction to N-mixture models: Short course. EURING Technical Meeting, Athens, Georgia, USA (28 April 2013).

Post van der Burg, M., B. Bly, T. VerCauteren, and A. J. Tyre. 2011. Making better sense of monitoring data from low density species using a spatially explicit modelling approach. Journal of Applied Ecology 48:47-55.

Royle, J. A. 2004. N-Mixture Models for Estimating Population Size from Spatially Replicated Counts. Biometrics 60: 108-115.

Royle, J. A., M. Kéry, R. Gautier, and H. Schmid. 2007a. Hierarchical spatial models of abundance and occurrence from imperfect survey data. Ecological Monographs 77:465-481.

Royle, J. A., R. M. Dorazio, and W. A. Link. 2007b. Analysis of multinomial models with unknown index using data augmentation. Journal of Computational and Graphical Statistics 16:67-85.

Royle, J. A., and R. M. Dorazio. 2008. Hierarchical modeling and inference in ecology: the analysis of data from populations, metapopulations and communities. Academic Press.

For more information on topics in this chapter

McKenney, H, T. M. Donovan, and J. Hines. 2007. Repeated count model (Royle). In Donovan, T. M. and J. Hines, eds. Exercises in occupancy modeling and estimation. Online: http://www.uvm.edu/envnr/vtcfwru/spreadsheets/occupancy.htm

Citing this primer

Powell, L. A., and G. A. Gale. 2015. Estimation of Parameters for Animal Populations: a primer for the rest of us. Caught Napping Publications: Lincoln, NE.

A biologist conducts a visual and auditory survey for songbirds in a grassland in northern Nebraska, USA. Photo by Silka Kempema, University of Nebraska-Lincoln.

Ch. 19

Distance sampling[1]

"There is no object so large but that at a great distance from the eye it does not appear smaller than a smaller object near."
-- Leonardo da Vinci

Questions to ponder:
- *Why is it necessary to record the distance to the animal during a survey?*
- *How can I estimate the effective strip width for a distance-based survey?*
- *What is a detection function?*
- *When should I use a point-count design, and when should I use line transects?*

Disappearing act

Ecologists often use visual surveys to count animals. And, as we explored in Chapter 14, there are many reasons to suspect that a count (c) does not equal the number of animals in a given area (N). Thus, we are usually very certain that $c \neq N$.

Distance sampling is a form of data collection intended for a specific type of analysis that uses detection functions to model the probability of detecting an animal, given a distance away from the transect or point-count location. Distance sampling is based on a basic concept—the probability of detecting an animal decreases as its distance from the observer increases. This concept could almost be referred to as a natural law! That is, it is difficult to imagine a situation in which our ability to detect an animal does not decrease at some point away from a transect. Even an elephant will eventually be too small to be seen!

When ecologists record survey data using distance sampling techniques, we see—over and over—that this "law" holds. Bigger animals may be seen for longer distances, but eventually even the number of large animals declines with distance away from the surveyor (Figure 19.1).

The approach for data collection when using distance sampling along a transect is to (1) establish the transect, (2) walk along the transect looking or listening for animals (or animal sign, e.g., dung or nests), and (3) recording the perpendicular distance from the animal to the transect line (x in Figure 19.2).

---

[1] *With thanks for content to Gary White and Michael Conroy*

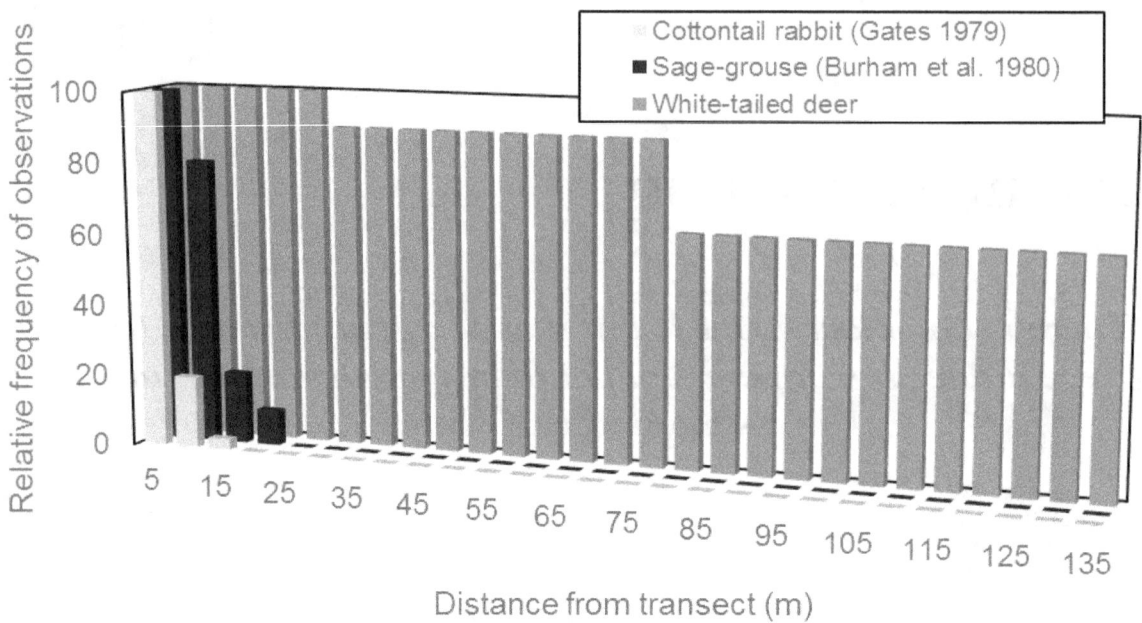

Figure 19.1: *Relative frequency of observations at given distances from a transect line for three species of wildlife during distance-based sampling: eastern cottontail (Sylvilagus floridanus) rabbits, greater sage-grouse (Centrocercus urophasianus), and white-tailed deer (Odocoileus virginianus). Figure after White 2007.*

A person may record the actual perpendicular distance (x) if that is possible, or the observer may record the radial distance to the animal (r) and the angle between the compass direction of the animal and the direction of the transect line (here, θ). Trigonometry can then be used to determine the perpendicular distance (x). Here, $\sin(\theta) = x/r$ so $x = \sin(\theta)/r$.

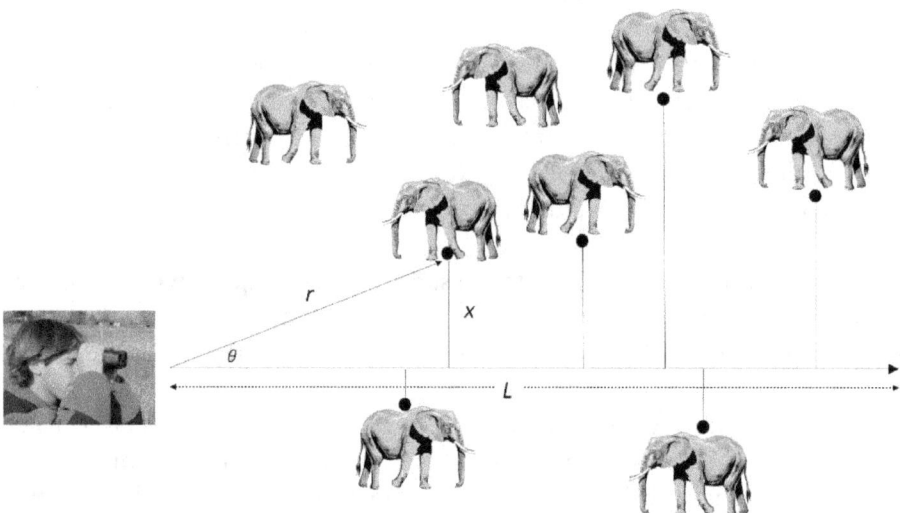

Figure 19.2: *Schematic showing perpendicular distance (x) recorded during a distance-based survey of detected (with distance) and undetected animals along a transect of length L.*

Building blocks

Distance sampling theory uses the set of perpendicular distances to estimate the density (D), or the number (N) of animals in a given area (a). We know that

$$D = \frac{N}{a}$$

But, we also know that **our count of animals is a sample**. Thus, we use n to represent the number of animals in our counted sample. And, if we look at a summary of our data, with observations binned into distance categories, we might see a pattern that looks something like the pattern in Figure 19.3b) at right. Simply put, it appears as if we have missed some animals at the farthest distances from the transect line. And, we can model that decline as a function—imagine placing a line through the tops of the bars in the histogram in Figure 19.3b. You would get the function shown in Figure 19.3c.

The key to understanding distance sampling is to understand that the area under the curve represents the size of the sample of animals that you saw (or heard). The area above the curve (Figure 19.3c, shaded) represents the number of animals that you missed—the animals that were present, but not detected. Obviously, if we do not correct for the missed animals, our density estimate is going to be biased (underestimated).

The correction provided by distance sampling is the probability of detection (here, $\hat{P}$):

$$\hat{D} = \frac{n}{a\hat{P}}$$

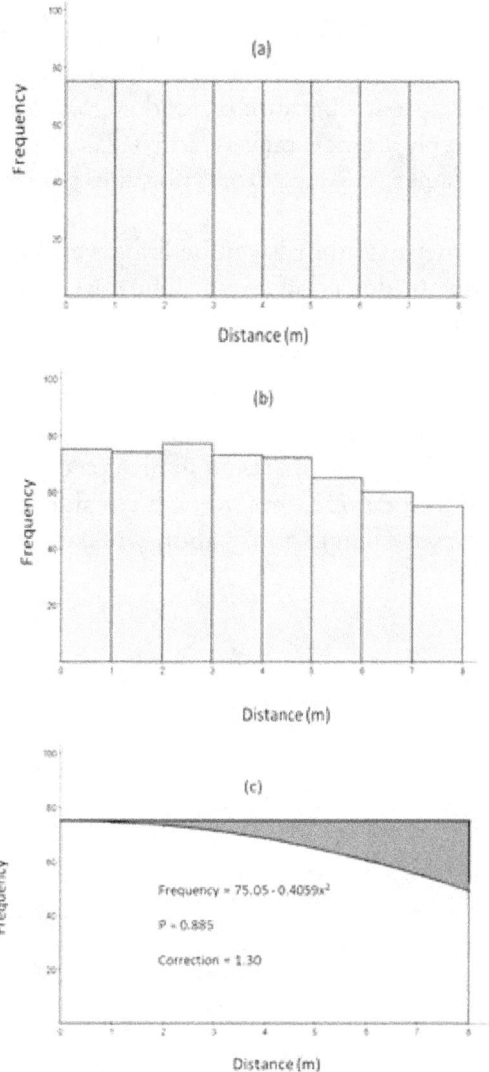

Figure 19.3: *survey results, in binned distance-from-line categories, expected with complete detectability (a) and incomplete detectability (b), as well as a detection function designed to describe the decline in detection as distance from line increases (c). Figure is after Buckland et al. 2001.*

Assumptions

The important **assumptions** inherent in analyses of distance-based data are as follows:

- Transect lines (or points) are randomly placed
- Objects directly on/at the line/point are detected with certainty (P = 1.0)
- Objects are detected at their initial location
- Measurements or groupings of data are correct
- Sightings of individuals are independent events

We assume random distribution of animals in space, relative to the placement of the transect line. Road surveys, for example, become problematic for distance surveys if animals are attracted to roads because of road edge habitats or provisioning from passing motorists (alternatively animals may avoid roads). Typically, it is better if transects are placed perpendicular (cutting across) possible gradients of habitat.

We assume that animals on the line or at the point are always detected, as we need a reference point for our detection curve. The curve for the probability of detection must start at 1.0 and decline in some manner. It is important for observers to collect exact measures, without error, because the estimation of the detection function is based upon this set of observations. As we will see, distance-sampling methods typically 'bin' or categorize distance data into groups, so the key is that animal observations are eventually placed into the correct distance category. Additionally, if surveys result in groups of animals (e.g., a covey of quail or a herd of deer) being detected at the same time, we will consider each group detection to be a single event. The user may indicate a single observation while also recording the size of the **cluster** of animals that was detected.

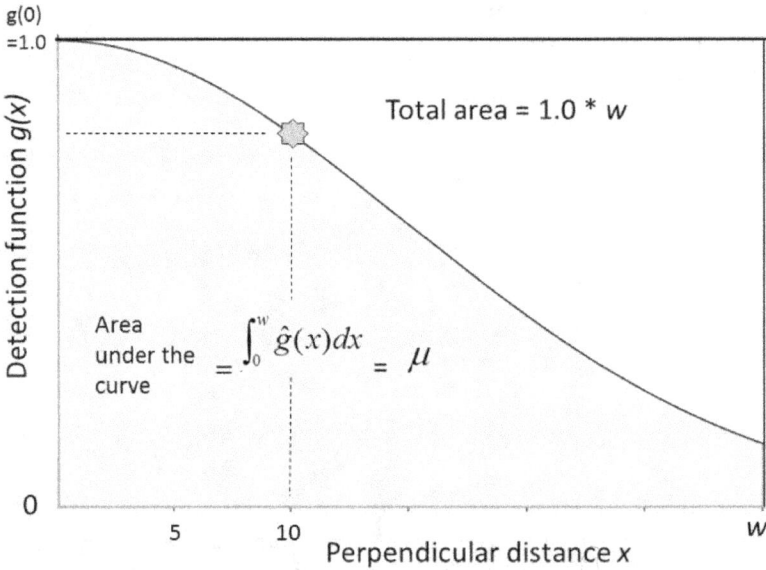

Figure 19.4: *Depiction of the detection function, g(x), declining from 1.0 to a minimum at, w, a distance from the transect. The value of g(10), or the detection probability at a distance of 10 units, is shown by dotted lines. Figure modified from Buckland et al. (2001).*

We can get a bit more specific in our analysis of the theoretical diagram (Figure 19.3c) for our detection function. First, we can name the function (the line) that describes the decline in probability of detection, g(x). Using this notation, we can refer to the probability at a certain distance (e.g., 10 meters) from the transect line as "g(10)", and we can see that the function would have a value (somewhere between 0 and 1.0) at the exact spot where x=10 in Figure 19.4. Second, if we imagine that our distances away from the line increase to some maximum distance, w, then we can also notice that the area of the entire rectangle (above and below the g(x) function) is 1.0*w—the rectangle has a height of 1.0 and a length of w. So, area=1.0*w = w.

Last, if we go back to our calculus course (perhaps you had no idea how useful that course would be when you took it, right!?), we know that it is possible to use the method of **integration** to estimate the area under the curve. Integration is a way to sum up the area under the curve by imagining a bazillion (yes, a bazillion) very skinny (in fact, infinitely small) rectangles packed together under the curve. Each rectangle, at distance x, has a height that matches the value of the function, g(x). So, we integrate them together with this formula,

$$\int_0^w g(x)dx = \mu$$

And we will use μ to represent this integration. So, the area under the curve equals μ. *That will be important.*

Conceptually, if the area under the curve represents the animals you saw, and the area of the rectangle represents all of the animals, we can estimate the detection probability, P, as the proportion of the rectangle that is under the curve:

$$P = \text{area under curve/area of rectangle}.$$

There is a theoretical problem with our goal to use the detection function to correct for the animals that we missed during our survey—**you literally cannot observe g(x), the detection function**. That is, we cannot observe and record probabilities, directly. But, we can record individual observations, and we can accumulate the frequency of our sightings away from the transect line as a **probability density function, f(x),** as in Figure 19.5. And that distribution of observations, f(x), can be used to estimate the number of animals that we missed. The area under the function f(x) is always scaled to equal 1.0, which makes f(x) a probability density function—in contrast to the detection function g(x) for which the area under the curve is much greater than 1.0. An

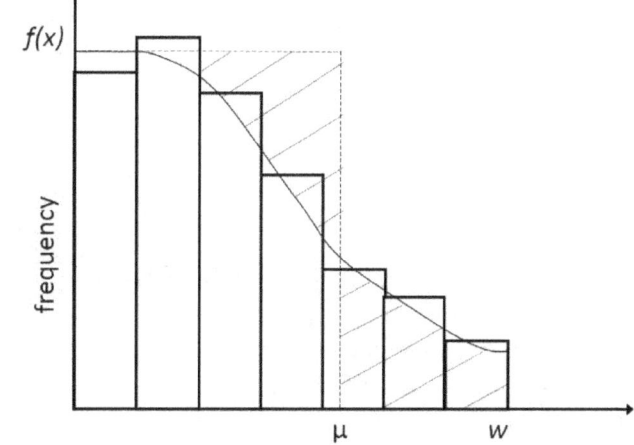

Figure 19.5: *The probability density function describing the distribution of perpendicular distance frequencies from a transect survey. The two shaded areas are equal in size so that the number of animals seen at distances >μ equals the number of animals not detected at distances <μ (figure after Buckland et al. 1998).*

example of a probability density function of which you may be familiar is the "bell curve", or normal distribution—100% of the students in a course fit under the function (the "curve").

As you can see in scenarios for line transects shown in Figures 19.4 and 19.5, line transect scenarios on previous pages, f(x), the distribution of observations, and g(x), the detection probability function, have the same shape. But, there are differences between f(x) and g(x). First, f(x) is built to describe how many animals are observed at distances away from the line— the function traces the histograms that summarize animal observations (Figure 19.6b), and the function is a **probability density function**.

In contrast, g(x) varies between 0 and 1.0 and describes the change in detection probability (Figures 19.4 and 19.6a). For point counts, the g(x) has the same general shape (or sometimes denoted g(r) for radial distances) as g(x) for transect counts (Figure 19.6a). But, f(x) (or f(r)) for point counts has a different shape (compare Figure 19.5 for transects and Figure 19.6b for point counts), because we expect fewer observations in the limited area near the point, because the center (e.g., 0-10 m from the point) of a 20-m radius circle has a smaller area than the area of the portion of the circle with radius 10-20 meters from the point. Detections increase with the area sampled away from the observer until some threshold distance whereupon detections begin to decline due to the effects of distance.

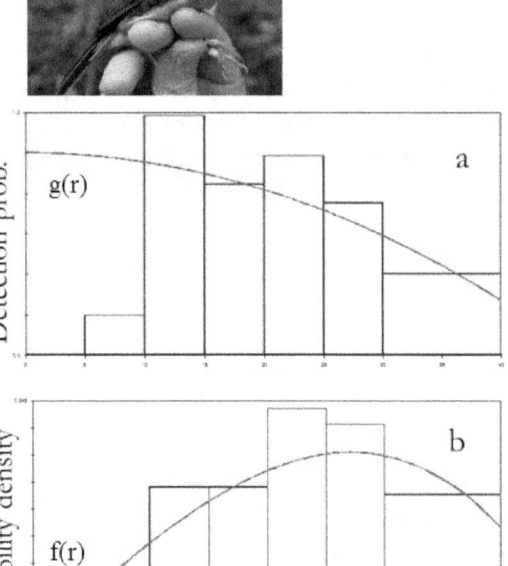

Figure 19.6: *Comparison of g(r), probability of detection at radial distance r (panel a), and f(r), the probability density function of observations at radial distances (panel b), for observations from point count surveys of hill blue flycatcher (Cyornis banyumas) in Thailand. Photo courtesy of G. Gale.*

The effective strip width concept

So, how do we estimate density if we don't know the area that we surveyed? Distance-sampling allows us to estimate the **effective area** that we surveyed, starting with the basic equation for density as the maximum likelihood estimator:

$$\hat{D} = \frac{N}{a}$$

And, as we previously noted, we are not able to count all N, and we are not completely certain about the area that we have counted, either. But, for a distance sampling research design with a transect of length L, and a theoretical width on either side of the transect, w, and an estimated detection probability, we can estimate density as:

$$\hat{D} = \frac{n}{2Lw\hat{P}}$$

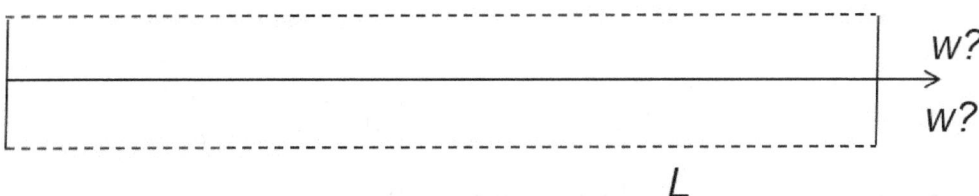

Figure 19.7: *Potential area surveyed along a line transect of length L and width, w. to each side of transect line.*

Distance sampling employs a concept described as the effective strip width, or more precisely effective "half-width." And, we should describe that to understand the logic of the method that distance sampling uses to estimate $\hat{D}$.

We start with the detection probability, P, which is the probability of detecting an animal in a strip surrounding a transect with a half-width (Figure 19.7) of w. Of course, w is unknown, as we do not need to enforce a fixed-width (or fixed-radius for point counts) rule when using distance sampling.

We have also already established that you can estimate P by dividing the area under the function, $g(x)$ by the area of the $1*w$ rectangle (and remember that we've established that μ is going to represent the integration of the area under the curve). Thus

$$\hat{P} = \frac{\int_0^w g(x)dx}{w} = \frac{\mu}{w}$$

So, if we replace P in our formula for density, we find:

$$\hat{D} = \frac{n}{2Lw\frac{\hat{\mu}}{w}} = \frac{n}{2L\hat{\mu}}$$

Through simplification, we see that μ ends up standing in the place of the "strip width" of the rectangle around the transect line (a=2Lw, now a=2Lμ). So, we can refer to μ as the effective strip width—the realized width on either side of our transect line after we account for our detection probability (P).

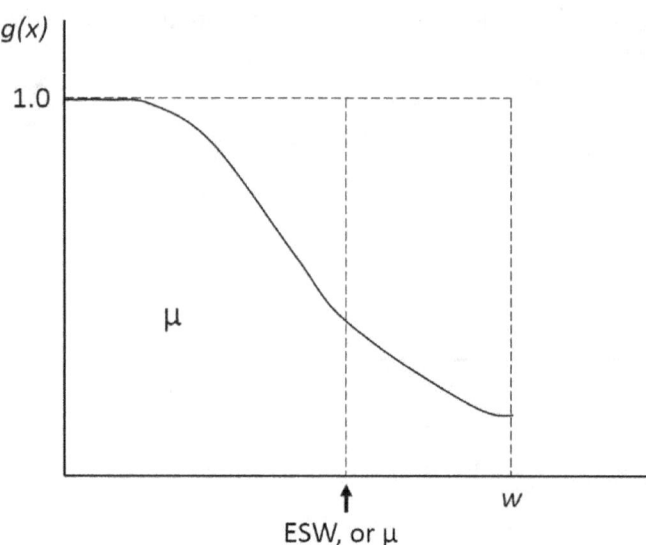

Figure 19.8: *The concept of μ: first, μ is the area under the detection function g(x) and takes a value between 0 and 1.0*w=w; second, μ is also the effective strip width, and takes a linear value between 0 and w (figure after Buckland et al. 1998).*

To visualize how μ, the effective strip width, can be seen as a distance <u>and</u> an area, consider Figure 19.8. We can define the effective strip width, μ, as the distance at which the number of animals missed within that distance is equal to the number of animals detected beyond that point. So, if we split our function, g(x) at the ESW, or μ, the area under the curve should equal the area above the curve for the other half of g(x).

Now, let's summarize the logic of distance sampling:

- We don't know the width of our transect (*w*) because of incomplete detection.
- We <u>do</u> know the distribution of our observations.
- We can fit a curve to observations that estimates probabilities of detecting at any distance, *x*. We can create a probability density function.
- That allows us to find ESW, or μ.
- We can use μ as our "w" to estimate D, because μ represents the width at which we have a count that is equivalent to of all animals in our sample (equal to our raw count).

The detection function

It should now be clear that the shape of the detection function is going to be important, as it will affect the effective half-width of the strip. And, although program Distance (the primary program used for distance sampling analyses) provides short-cuts to create the linear models used in the analysis, those equations are important to using and interpreting results from distance sampling.

Distance allows the fitting of complex detection functions of the form:

$$g(x) = key(x)[1 + series(x)]$$

Each function is based on a **key function** that determines the general shape of the function.

The key functions used by Distance are (Figure 19.9):

- **Half normal**—describes a gentle decline in probability of detection that follows the well-known 'bell curve' shape
- **Uniform**—describes a scenario in which probability of detection does not fall
- **Negative exponential**—describes an immediate drop in probability of detection at a very fast rate (an exponential rate!)
- **Hazard**—describes a scenario in which the probability of detection remains high until a threshold point, at which detection drops precipitously.

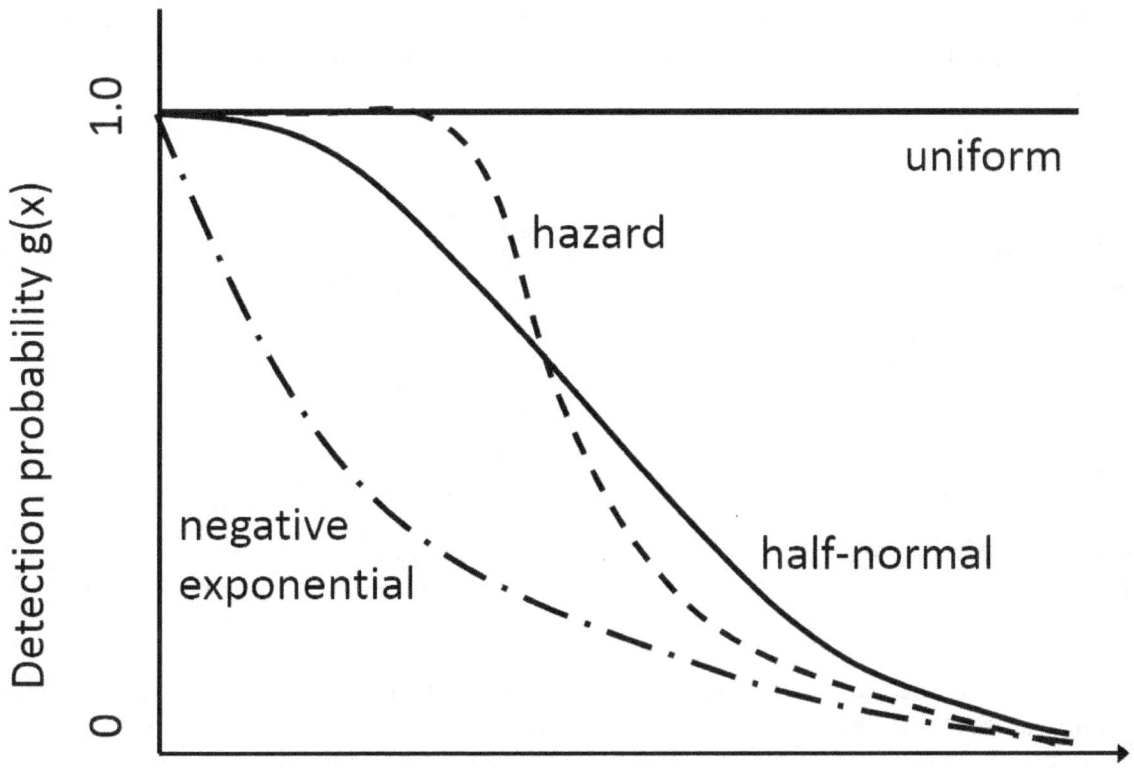

Figure 19.9: *General shapes of 4 key functions used by program Distance to describe the decline in detection as a function of distance from the transect line or point.*

Program Distance uses the four basic key functions, or shapes (Figure 19.9), because they are generally flexible shapes with fairly simple mathematical structure (and thus have low variance), and each function meets the criteria that we must have $g(0) = 1$. That is, the functions can all have some type of "shoulder" at $x = 0$ before declining as distance increases away from the transect line.

Then, Distance uses a **series adjustment** that provides a "tweak" to the basic shape of the function—especially at the far end. These adjustments (Figure 19.10) allow the g(x) function to better fit the data. However, the adjustments also add parameters to the function, and thus make it more complex—so, we use AIC to assess the models.

- Cosine
- Simple polynomial
- Hermite polynomial

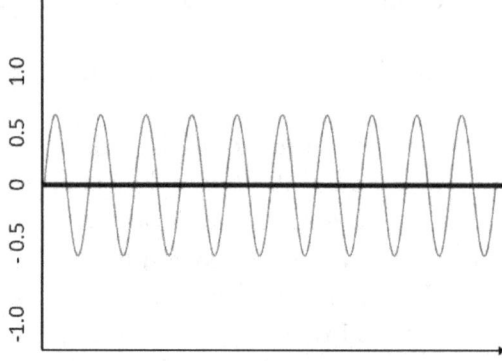

Figure 19.10: *An example of a cosine function used as a series adjustment to key functions by Distance.*

We can look at an example of how a key function can be modified to provide better fit to distance-based survey data. For example, if we were to try to use a uniform key function without any adjustments (Figure 19.11a), our eyes tell us that it does not do a good job at describing the manner in which the probability of detection appears to decline with distance away from the transect line.

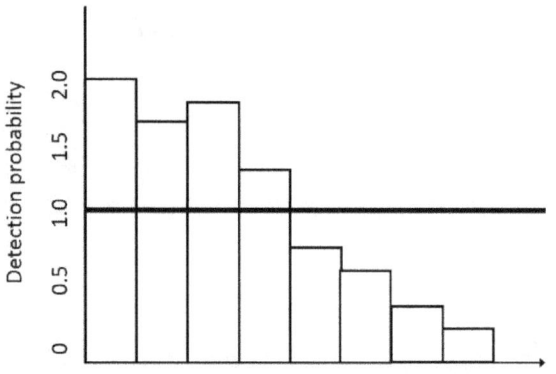

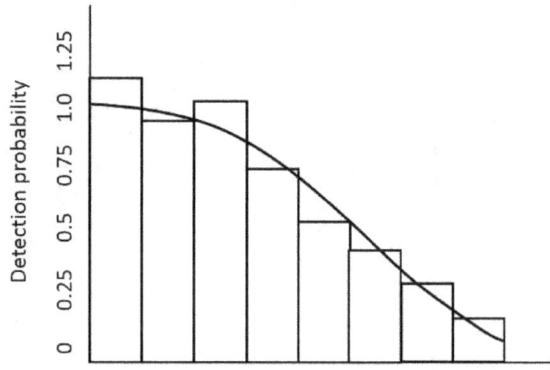

Figure 19.11a: *A uniform distribution fitted to survey observations (Figure after White 2007).*

Figure 19.11b: *Observations from a survey fitted with a uniform key function with a cosine adjustment term (after White 2007).*

But, we can ask the software to add one or more cosine adjustment terms to the linear model describing the shape of the detection function. And, for example, when we add one cosine adjustment term (Figure 19.10) to the uniform key function, we get an adjusted uniform detection function (Figure 19.11b).

The adjusted function (Figure 19.11b) appears to fit the distribution of our observations much better. And, we note that it looks somewhat similar to the half-normal function. However, the detection probability does not fall towards 0 as quickly as the half normal would. For the other adjustment terms, the simple polynomial and hermite polynomial, the adjustments are applied similarly to cosine function except using a polynomial or a hermite polynomial series adjustment. The two polynomial adjustments often produce identical results, but the adjusted functions will differ in the amount and gradualness of the curvature relative to the cosine adjustment depending on the data being fitted.

Step-down model selection

Program Distance (Thomas et al. 2010) provides a robust example of the **step-down model selection** process that we discussed in Chapter 4. The purpose of this process is to discard models with uninformative parameters. Before exploring whether density of animals varies between geographic strata or other groupings, Distance compares the competing models (*key functions*) that describe the decline in detection probability. In fact, Distance starts the selection process with an examination of the best way to "tweak" the key function—as we saw above with the cosine adjustment of the uniform key function.

Thus, Distance first uses AIC (or AIC_c, see Chapter 4) to select the best adjustment term for one or more key functions. Then, the second step is to take the best form of each key function (now modified with an adjustment term) to compare with one or more of the other key functions. Now, Distance uses a second AIC assessment to compare the half-normal with, for example, the hazard and the uniform key functions. And, the last step is to compare the models that represent various hypotheses about density—using the best description of detection probability.

Distance does not evaluate the density hypotheses by comparing all key functions with each possible adjustment term in a simultaneous model comparison. *No, Distance discards uninformative models along the three-step process.* It is also important to remember that AIC cannot be used to compare between different datasets. For example, AIC cannot be used to compare between models if the data are grouped or truncated differently. In such cases the coefficient of variation is useful for deciding where to truncate or how to group the data before proceeding with model selection.

Point-counts

To this stage, we've mostly discussed the application of distance sampling to transect surveys. But, many people use point counts to survey for birds or butterflies or other species. The theory of distance sampling can be applied to point-counts as well. Some research biologists will refer to such efforts as **variable circular plots** (Figure 19.12), meaning that there is no fixed radius used for the survey. If an animal is seen or heard, the

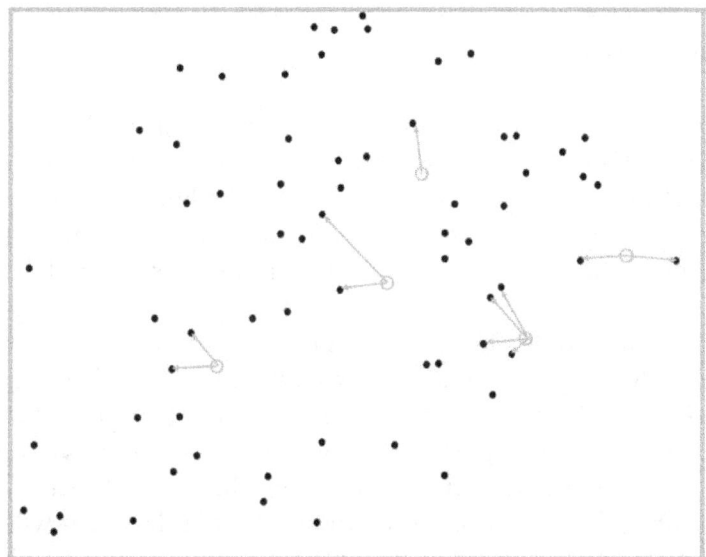

Figure 19.12: *Depiction of 5 point counts conducted on a study site with 11 observed animals; relative radial distances are shown (Figure modified from Buckland et al. 2001).*

distance to that animal is recorded. Thus, our data will consist of sets of radial distances from the point to the animal.

Transects vs. point-counts

When should you use line transects and when should you use point-counts? We can provide some general suggestions for your consideration as you design a study.

Line transects are generally best for:

- Sparsely distributed populations for which sampling needs to be efficient (e.g., whales, deer, grassland birds). You might consider using line transects if you think that you will often not count any individuals at a given point-count location.
- Populations that occur in well-defined clusters, and at low or medium density.
- Populations that are detected through a flushing response, where the act of walking the transect helps flush the animals.

Point-counts are generally best for:

- Patchily distributed populations, as you can easily stratify points to patches.
- Populations that occur in difficult terrain, or with problematic access—walking a straight transect as you survey for animals in these locations may be hazardous to your health if you cannot keep track of where you should be stepping!
- Associating survey data with associated habitat characteristics—a point is located at a discrete location at which you can measure vegetation. Associating vegetation with transects is more difficult, as transects are continuous.
- Populations that occur at medium or high densities—points are not as effective at low densities, because you obtain very few observations from each point.

Details to consider

Not everyone's data will follow a nice continuously declining function. In the songbird example depicted in Figure 19.13a, you can see that the surveyors in Nebraska detected western meadowlarks a considerable distance from the line. But, there were also gaps where animals were not seen at some distance categories. If you try to imagine fitting one of the basic key functions to this data, you realize it may not be easy—the fit may be poor.

More critically, these outlier observations can affect the shape of the detection functions fairly dramatically. **Right truncation** is recommended (Buckland et al. 2001), which means that the outliers should be removed before analysis begins (Figure 19.13). Although some ecologists use a rule, such as removing 5% of the observations to remove the outliers, there are good reasons to use a **visual assessment** to determine which outliers that should be removed. It is possible, for example, that a portion of the top 5% of the distances will not be in the 'outlier' group, and thus you will have removed some of your sample for no good reason. *We always recommend looking at summaries of your data before jumping into the analyses.*

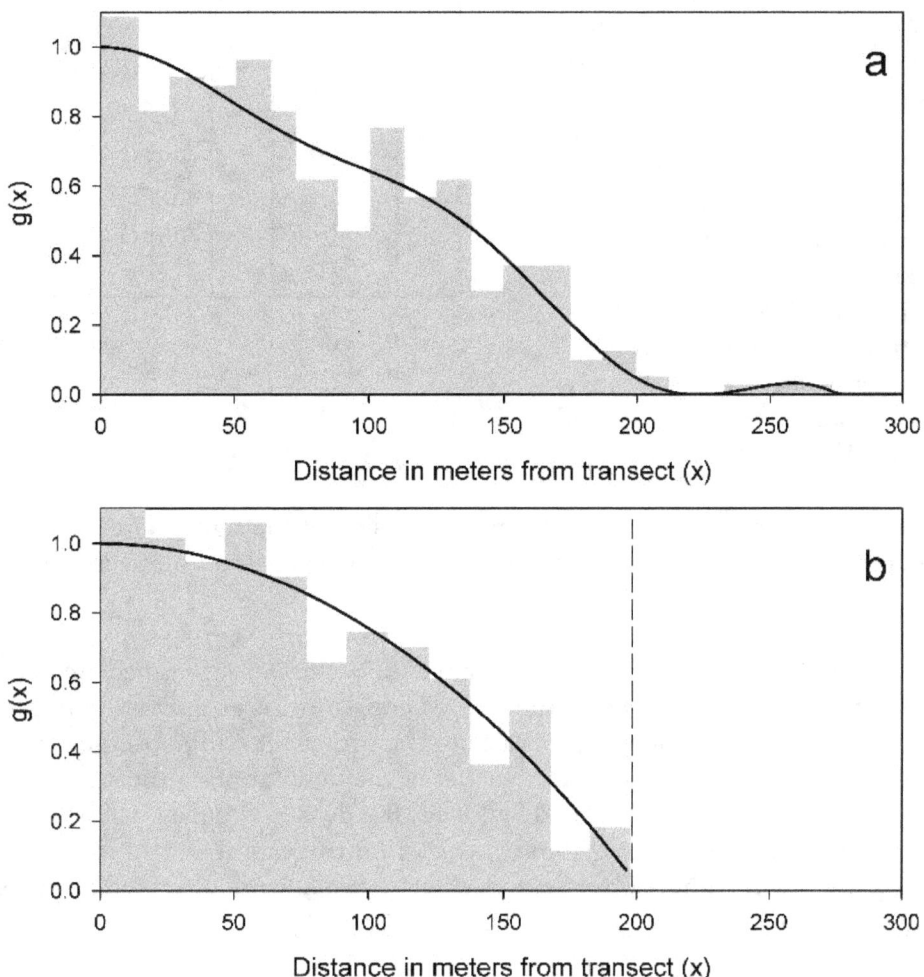

Figure 19.13: *Detection functions, g(x), for observations of western meadowlarks (Sturnella neglecta) from line transects in grasslands in Nebraska, USA (Kempema 2007). The full data set is shown at top (a), and the right-truncated data set is shown below (b, the data has also been grouped into larger bins).*

If right truncation is a good thing, perhaps **left truncation** is also a good thing? *We suggest not.*

Some ecologists might be tempted to use left truncation when working with point-counts. One potential problem that is encountered when using distance sampling and point-counts of grassland birds for example, is that the observer can create a "halo effect" (inset, Figure 19.14). As the observer walks to the point-count location, birds may move away from the point. Then, the distribution of the observations may not meet our expectations—we may have fewer observations that we expected near the 0-point (Figure 19.14).

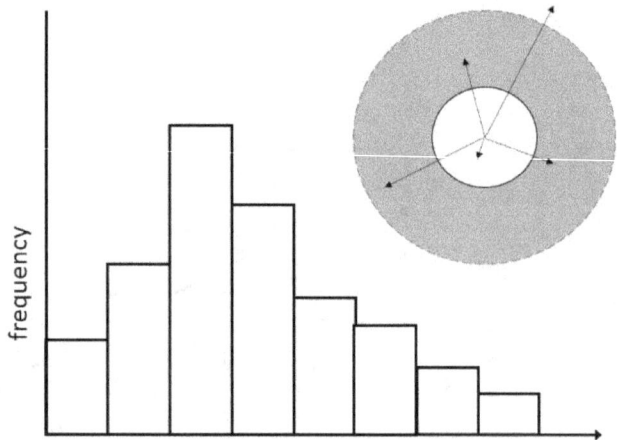

Figure 19.14: *Observations from a point-count survey, showing effects of the observer's presence at the point; inset, the effects of left truncation on the area included in the estimate of density.*

Left truncation removes the observations from the left side of the figure here. And it reduces the effective diameter of the circle by creating a 'doughnut' shape—a hole is created in the middle of the point count.

The problem with this approach is that you still have the observations immediately to the right of the left truncation zone. And, those observations consist of some animals that left the truncated portion of the point-count in response to the presence of the observer. As they moved outwards, they increased (biased) the density higher in the outer zones—there are now more animals in the outer zone than there should be. And, the density estimate will be biased high. For this reason, left truncation is not recommended. Left truncation is only advocated for a very limited set of circumstances where observations close to the line are obscured or problematic (see Buckland et al. 2001).

If you are faced with this problem, you may wish to consider using **N-mixture modeling** (Chapter 18) as an approach to compare relative abundance, rather than distance sampling. The N-mixture modeling design is just based on the count—the movement of the animals is not a problem in this type of study design.

Distance needs data: a lot of data

In our primer, we've tried to avoid explicitly making our discussions based on the software you might use to analyze your data. However, a great majority of people attempting a distance-based approach will use program Distance. When you receive your results from Distance, you will be given a table of parameter estimates. These parameters are mostly "nuisance" parameters—parameters that had to be estimated to allow estimation of density ($\hat{D}$).

Here are the parameter estimates from a fox squirrel (*Sciurus niger*) survey in Lincoln, Nebraska, USA by University of Nebraska-Lincoln students. Students conducted point-counts on campus.

The model selected through step-down model selection was the hazard rate key function with no adjustment terms:

```
    Effort          :    58.00000
    # samples       :    58
    Width           :    169.0000
    # observations  :    67

Model
   Hazard Rate key, k(y) = 1 - Exp(-(y/A(1))**-A(2))
```

Parameter	Point Estimate	Standard Error	Percent Coef. of Variation	95 Percent Confidence Interval	
A(1)	30.97	4.493			
A(2)	3.350	0.4096			
h(0)	0.001016	0.000203	19.99	0.000684	0.001509
p	0.068902	0.013776	19.99	0.046400	0.102320
EDR	44.361	4.4346	10.00	36.351	54.137

We show you these results to impress upon you that distance-sampling requires the estimation of many parameters—not just density. Here, *A(1)* and *A(2)* are parameters used in the non-linear model (hazard function) for probability of detection (Figure 19.15). And, this key function has only two parameters—if adjustment terms had been selected, we would see parameter estimates for the parameters used to adjust the key function, which would make our model even more complex.

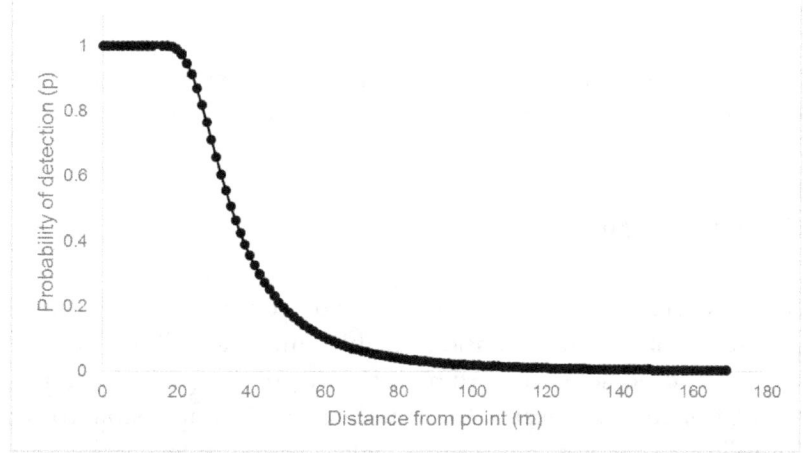

Figure 19.15: *Detection curve (probability of detection, g(x)) for a fox squirrel survey conducted in September 2014 by students at the University of Nebraska-Lincoln, USA. The curve is shaped as a hazard function, which was selected as the model to describe the decline in detection as distance increased from the point count location (goodness of fit: $X^2_2 = 0.21$, $P = 0.90$).*

Then, we see three parameters, *h(0)* (a measure of the effective detection area), *p* (probability of detection for the study), and EDR (the effective detection radius for the squirrel point counts) that can be estimated after f(x) and g(x) are established. Then, as a third step, density is estimated using EDR and *p*. The probability of detection for animals near points was low, and the detection curve (Figure 19.15) declined much more rapidly than students anticipated before conducting the survey—trees, shrubs, and buildings on campus most likely combined with the small size and camouflage of the squirrel such that students had less than a 20% chance of seeing a squirrel that was more than 40 meters from the survey point. In our example, the density for fox squirrels, per hectare, was estimated as:

	Estimate	%CV	df	95% Confidence Interval	
D	1.8685	20.68	73.70	1.2427	2.8094

It should now be clear that our estimate for density, and its variance, are both a function of several other parameters. If you do not have enough observations for a species, it will be difficult to find a well-fitting description of the decline in detection probability away from your lines or points. The uncertainty in the key function for detection will manifest itself as high variance associated with your detection probability, p. And, high variance in p will result in high variance for D.

Therefore, study design and forethought are critical for distance-sampling. The University of Nebraska squirrel survey worked well to provide a reasonably precise estimate of density. But, you must plan to conduct enough effort to collect adequate data (see Buckland et al. 2001 for advice on planning). In general "adequate" data entirely depends on the quality of the detection function- if the data are messy even a sample of 120 might not be large enough but on the other hand a well-formed detection function might give a reasonable estimate with as few as 30 detections.

Distance-sampling has one clear advantage over all other types of survey design and analysis—**it is the only method to provide *bona fide* (direct) density estimates**. All other methods estimate abundance (N), which may be transformed to density (D) estimates only after making some rather important, and often very ill-informed, assumptions about the area from which the estimate of N was made.

So, the trade-off for the use of distance sampling is between amount of data needed for the parameter estimates and the precision and bias of the density estimate provided. You must evaluate this tradeoff for your research questions.

Availability...?

Surveys used to estimate density should account for detectability (the probability that an animal is detected if it is present and available) and availability (the probability that the animal is available to be detected). Animals below ground (see Chapters 13 and 14) or quietly hidden in vegetation are unavailable for sampling, as it is impossible to see or hear them. Distance sampling, in its most basic form, only deals with detectability, and most users of distance sampling typically assume an availability of 1.0. This is often unrealistic (Diefenbach et al. 2007), but clearly depends on the species of interest. Availability for some species is often as low as 0.1, and unadjusted distance sampling may underestimate density by as much as a factor of 4. We refer the reader to discussion of a number of methods to account for availability of singing/calling birds by Diefenbach et al. (2007).

Conclusion

Distance sampling uses a maximum-likelihood estimation process to estimate density. The method addresses a basic "law" regarding observations during surveys—the probability of detection for an animal declines the farther the animal is from the observer. Distance sampling is rigorous and robust if assumptions are met. In its simplest form, distance sampling does not provide an estimate of the probability of availability, but it deals very effectively with detectability. Distance sampling can be applied to many species and types of surveys.

References

Buckland, S.T., D. R. Anderson, K. P. Burnham, and J. L. Laake. 1998. Distance sampling, in Encyclopedia of Biostatistics, P. Armitage and T. Colton, eds, John Wiley & Sons, Ltd, Chichester.

Buckland, S. T., D. R. Anderson, K. P. Burnham, J. L. Laake, D. L. Borchers, and L. Thomas. 2001. Introduction to Distance Sampling: Estimating Abundance of Biological Populations. Oxford University Press, New York.

Diefenbach, D. R., M. R. Marshall, J. A. Mattice, and D. W. Brauning. 2007. Incorporating availability for detection in estimates of bird abundance. Auk 124:96–106.

Kempema, S. L. F. 2007. The influence of grazing systems on grassland bird density, productivity, and species richness on private rangeland in the Nebraska Sandhills. MS Thesis, University of Nebraska-Lincoln, Lincoln, NE.

Thomas, L., S.T. Buckland, E.A. Rexstad, J. L. Laake, S. Strindberg, S. L. Hedley, J. R.B. Bishop, T. A. Marques, and K. P. Burnham. 2010. Distance software: design and analysis of distance sampling surveys for estimating population size. Journal of Applied Ecology 47: 5-14.

White, G. 2007. FW663 -- Sampling & Analysis of Vertebrate Populations. On-line: http://warnercnr.colostate.edu/~gwhite/fw663/ Accessed 15 February 2015.

For more information on topics in this chapter

Conroy, M. J., and J. P. Carroll. 2009. Quantitative Conservation of Vertebrates. Wiley-Blackwell: Sussex, UK.

Williams, B. K., J. D. Nichols, and M. J. Conroy. 2002. Analysis and management of animal populations. Academic Press, San Diego.

Citing this primer

Powell, L. A., and G. A. Gale. 2015. Estimation of Parameters for Animal Populations: a primer for the rest of us. Caught Napping Publications: Lincoln, NE.

Caught Napping Publications
Lincoln, Nebraska USA

www.ingramcontent.com/pod-product-compliance
Lightning Source LLC
Chambersburg PA
CBHW080907170526
45158CB00008B/2021